The United States Marine Corps in the World War

By

EDWIN N. McCLELLAN
Major, U. S. Marines
Officer in Charge Historical Division

WASHINGTON
GOVERNMENT PRINTING OFFICE
1920

EXPLANATORY NOTE.

This brief history has been prepared for the purpose of acquainting both the personnel of the service and the public with the general facts concerning the United States Marine Corps in the World War.

It is a partial compliance with the instructions contained in Marine Corps Orders No. 53 (Series 1919), directing that a history of the United States Marine Corps for the period of the World War be prepared, and is preliminary to the final and detailed history of the United States Marine Corps during the World War, which is in course of preparation.

The statistics and other information contained herein are as accurate as it is possible to obtain at the present date. Every effort has been made to avoid expressions of opinions and criticisms, or the drawing of conclusions of an important nature.

LETTER OF TRANSMISSION.

NOVEMBER 26, 1919.

From: Officer in Charge Historical Division, Adjutant and Inspector's Department, United States Marine Corps.
To: The Major General Commandant.
Via: Officer in Charge, Adjutant and Inspector's Department.
Subject: The United States Marine Corps in the World War.

1. There is transmitted herewith for your formal approval a concise history of the United States Marine Corps in the World War, including certain statistics, with the recommendation that it be published to the naval service.

EDWIN N. MCCLELLAN.

[First endorsement.]

ADJUTANT AND INSPECTOR'S DEPARTMENT,
HEADQUARTERS UNITED STATES MARINE CORPS,
Washington, D. C., November 26, 1919.

From: The Acting Adjutant and Inspector.
To: The Major General Commandant.

1. Forwarded, approved.

H. C. HAINES.

Approved:
GEORGE BARNETT,
Major General Commandant,
United States Marine Corps.

Approved:
JOSEPHUS DANIELS,
Secretary of the Navy.

TABLE OF CONTENTS.

THE UNITED STATES MARINE CORPS IN THE WORLD WAR.

Chapter I.

IN GENERAL.

When a state of war was declared to exist on April 6, 1917, the United States Marine Corps was composed of 462 commissioned officers, 49 warrant officers, and 13,214 enlisted men on active duty, a total of 13,725 and, while the corps was expanded to an actual strength, including reserves, of 75,101 officers and enlisted men, its high standard was never lowered. When these figures are compared with the approximate strength of 3,100 at the end of the Civil War, and of 4,800 at the end of the Spanish War, the growth of the Marine Corps is illustrated.

Despite the fact that on the outbreak of war, 187 officers and 4,546 enlisted men were on duty beyond the continental limits of the United States, and 49 officers, and 2,187 enlisted men were serving on board the cruising vessels of the Navy, only five weeks later, on June 14, 1917, the Fifth Regiment of Marines, consisting of 70 officers and 2,689 enlisted men, approximately one-sixth of the enlisted strength of the Marine Corps, competently organized and ready for active service, sailed on the *Henderson, De Kalb*, and *Hancock* from the United States, forming one-fifth of the first expedition of American troops for service in France.

This regiment was soon joined by the Sixth Regiment and the Sixth Machine Gun Battalion of Marines, and the Fourth Brigade of Marines was organized, and as one of the two Infantry brigades of the Second Division of Regulars engaged in actual battle in no less than eight distinct operations in France, of which four were major operations.

The French Army recognized the splendid work of the Fifth and Sixth Regiments of Marines by citing them no less than three times in Army orders for achievements in the Chateau-Thierry sector, in the Aisne-Marne (Soissons) offensive, and in the Meuse-Argonne (Champagne). The Sixth Machine Gun Battalion was similarly cited for its work in the Chateau-Thierry sector and Aisne-Marne (Soissons) offensive. The Fourth Brigade received a similar citation for its work in the Chateau-Thierry sector. Since two French Army citations are sufficient to make an organization eligible for the award of the French fourragére, the high standard of the Marine units is evident. Information was received in January, 1920, that the War Department had accepted the award of the French fourragére in the colors of the ribbon of the Croix de Guerre for several Army organizations and the three units of the Fourth Brigade.

Within one year after the outbreak of war the Marine Corps placed about as many enlisted men in France as there were in the Marine Corps when war was declared.

During the month of June, 1918, when the battle deaths around Hill 142, Boureschcs, Belleau Wood, and Vaux, of Americans attached to the Second Division amounted to 1,811 (of which at least 1,062 were Marines) and the nonfatal casualties to 7,252 more (of which 3,615 were Marines), the legislative strength of the Marine Corps was but 1,323 officers and 30,000 enlisted men; the actual strength on June 30, 1918, including reserves, was 1,424 officers and 57,298 enlisted men, and of this total about 300 officers and 14,000 enlisted men were in France. These latter figures include those Marines who suffered casualties in the battles of June, 1918.

Approximately 30,000 Marines were sent overseas to join the American Expeditionary Forces, and 1,600 for naval duty ashore.

During the war a great many additional Marine detachments were detailed to guard the radio stations, naval magazines, ammunition depots, warehouses, cable stations and for other naval activities, and the detachments already established were largely augmented. No call was made for additional Marines for naval purposes that was not fully met, and this is of especial interest as the Marine Corps is essentially a part of the Naval Establishment, and its first duty is to fill all naval needs and requirements. It was believed to be essential that the Marine Corps should do its full part in this war, and for that reason it was absolutely necessary that the Marines should join the Army on the western front, taking care, however, that this should not at any time interfere in the slightest degree with the filling of all naval requirements.

The Marine Corps, while maintaining the Fourth Brigade of Marines, a total of 258 officers and 8,211 enlisted men, that fought in eight battle operations suffering approximately 12,000 casualties, placed and maintained the Fifth Brigade of Marines of the same strength in France; supplied the commanding general of the Second Division, and many officers on his staff; furnished a considerable number of officers to command Army units of the Second and other divisions, and for staff and detached duty throughout the American Expeditionary Forces; participated in the naval aviation activities in France and in the Azores; and during the period of the war succeeded in performing in a highly satisfactory manner the naval duties required of it, including the maintenance of two brigades of prewar strength standing by to protect the Mexican oil fields, and as an advanced base force in Philadelphia; one in Cuba; one in Santo Domingo, and one in Haiti; administered and officered the Haitian Gendarmerie and Guardia Nacional Dominicana; as well as providing efficient Marine detachments for numerous naval vessels, and maintaining garrisons at the numerous navy yards and naval stations in the United States; and in the Virgin Islands; Guantanamo Bay, Cuba; Pearl Harbor, Hawaiian Islands; Guam; Cavite and Olongapo, P. I.; Managua, Nicaragua; Peking, China; San Juan, P. R.; London, England; Cardiff, Wales; Paris, France; and the Azores; and supplied many officers and enlisted men for special and detached duty at home and abroad.

Chapter II.

STATUTORY AND ACTUAL STRENGTH OF THE MARINE CORPS ON VARIOUS DATES.

STATUTORY STRENGTH.

The act of Congress of August 29, 1916, increased the authorized strength of the Marine Corps from 344 officers and 9,921 enlisted men to 597 officers and 14,981 enlisted men, and the President was authorized in an emergency to further increase the corps to 693 officers and 17,400 enlisted men, which he did by Executive order on March 26, 1917.

On April 6, 1917, Congress declared "that a state of war exists between the United States and the Imperial German Government" and one and one-half months later, on May 22, 1917, temporarily increased the authorized strength to 1,197 commissioned officers, 126 warrant officers, and 30,000 enlisted men. Finally, the act of July 1, 1918, temporarily increased the Marine Corps to 3,017 commissioned officers, 324 warrant officers, and 75,500 enlisted men, which is the maximum strength ever authorized for the Marine Corps. Of this number 17,400 were permanent and 57,650 temporary. In addition to the above, the act of August 29, 1916, which established the Marine Corps Reserve, permits the enrollment of reserves without limit as to number, and on April 6, 1917, there were enrolled, subject to call to active duty, three Reserve commissioned officers, 24 National Naval Volunteer officers, 36 Reserve enlisted men, and 928 enlisted National Naval Volunteers. There were also available for recall to active duty 65 regular retired commissioned officers, one regular retired warrant officer, and 210 regular retired enlisted men.

ACTUAL STRENGTH OF THE MARINE CORPS AT THE BEGINNING AND END OF THE WAR.

On April 6, 1917, the strength of the Marine Corps on active duty was as follows:

Regular commissioned officers:	
Major General Commandant	1
Brigadier generals	7
Colonels	13
Lieutenant colonels	27
Majors	59
Captains	119
First lieutenants	87
Second lieutenants	106
Total regular officers	419
Regular commissioned retired officers:	
On active duty	43

Regular warrant officers:	
Marine gunners	20
Quartermaster clerks	20
Pay clerks	9
Total warrant officers	49
Total regular officers	511
Total regular enlisted men	13, 214
Total strength on active duty	13, 725

On November 11, 1918, the strength of the Marine Corps on active duty was as follows:

Regular commissioned officers:	
Major General Commandant	1
Major generals	2
Brigadier generals	13
Colonels	43
Lieutenant-colonels	52
Majors	199
Captains	522
First lieutenants	436
Second lieutenants	413
Total Regular officers	1, 681
Commissioned retired officers:	
On active duty	43
Reserve officers on active duty:	
Majors	7
Captains	33
First lieutenants	63
Second lieutenants	360
Total Reserve officers	463
Total commissioned officers an active duty	2, 187
Regular warrant officers:	
Marine gunners	109
Quartermaster clerks	89
Pay clerks	56
Total	254
Reserve warrant officers:	
Marine gunners	27
Quartermaster clerks	2
Pay clerks	4
Total	33
Total warrant officers on active duty	287
Total officers on active duty	2, 474
Enlisted personnel:	
Regular	63, 714
Retired enlisted men on active duty	15
Reserves, on active duty	6, 483
Female reservists, on active duty	277
Total	70, 489
Total strength on active duty	72, 963

On December 11, 1918, the Marine Corps attained its maximum strength on active duty, which was distributed as follows:

Regular commissioned officers	1,678
Retired officers on active duty	44
Reserve commissioned officers	452
Regular warrant officers	257
Reserve warrant officers	31
Regular enlisted men	65,666
Reserve enlisted men	6,704
Female reservists	269
Total	75,101

The maximum enlisted strength of the regular Marine Corps, not including reserves, during the period between the outbreak of war and the date the armistice became operative was 63,714 on November 9, 1918.

Chapter III.

RECRUITING—APPLICANTS, REJECTIONS, ENLISTMENTS—ENLISTMENTS BY STATES.

The recruiting service of the corps was enlarged greatly during the war and it was so well organized and its method of procedure was so efficient that it was able to stand the enormous increase of the corps. The real test of any organization comes when a very great increase is suddenly made and the recruiting service of the Marine Corps passed that test in a commendable manner.

On August 8, 1918, by Executive order, volunteer enlistments in the Marine Corps and enrollments in the reserve were stopped, and from that time until October 1, 1918, no men were enlisted in the corps with the exception of those whose cases were pending when the Executive order above mentioned was issued and some whose enlistments expired and were reenlisted. On September 16, 1918, the Secretary of War approved the terms of a tentative plan proposed in an informal conference by representatives of the Navy Department, the Marine Corps, the General Staff, and the Provost Marshal General's Office.

This plan in part provided that the Marine Corps was accorded the privilege of individual inductions to the amount of 5,000 men, for the months of October, November, and December, 1918, and January, 1919, and 1,500 thereafter.

As the plan above mentioned operated the men were supplied from the selective draft, but the choice was given the Marine Corps of accepting or rejecting men according to the way they measured up to the Marine Corps standards. The inductees also had a choice in the matter, so they were really "voluntary inductees." This plan was very favorable and permitted the Marine Corps to maintain its high standard of enlisted personnel.

Owing to the cessation of hostilities there were but few inductions and none of the inductees ever reached France prior to the armistice becoming effective. Regular voluntary inductions into the Marine Corps (through Provost Marshal General) commenced October 1, 1918, and the last man was voluntarily inducted on December 13, 1918. Inductions occurred as follows:

October, 1918	2,787
November, 1918	3,880
December, 1918	421
Total	7,088

Owing to the signing of the armistice, no more requests were made to the Provost Marshal General for the induction of men after November 18, 1918.

On December 2, 1918, the President, by proclamation, directed that voluntary enlistments of registrants into the Navy and Marine

Corps would be permitted without notice to local boards, and the provisions of the selective service law became inoperative so far as the Marine Corps was concerned.

On December 4, 1918, recruiting on a very limited scale was resumed by order of the Secretary of the Navy. On that date also, enrollments in the Marine Corps Reserve were stopped.

Applicants, rejections, enlistments, etc., regular Marine Corps, not including reserves but including inductees, April, 1917, to November, 1918.

Date.	Applicants.	Rejected by commanding officer.[1]	Rejected by medical officer.[2]	Eloped.	Declined oath.	Enlistments.	Strength Marine Corps.
Apr. 1	....	....	....	....	....	....	13,214
Apr. 30	14,607	41	11,673	10	19	2,864	15,813
May 31	15,498	74	10,039	40	50	5,295	20,932
June 30	15,905	47	11,735	16	34	4,073	24,772
July 31	11,778	21	8,183	22	44	3,508	27,045
Aug. 31	6,275	37	4,006	7	4	2,221	29,861
Sept. 30	4,846	29	3,996	5	5	811	30,322
Oct. 31	4,335	33	3,661	5	1	635	30,576
Nov. 30	5,577	14	4,942	2	2	617	30,855
Dec. 31	6,788	22	5,305	4	5	1,452	32,016
Jan. 31	5,472	29	3,981	5	3	1,454	33,184
Feb. 28	5,915	31	5,772	4	3	105	33,045
Mar. 31	5,037	18	4,734	2	4	279	33,093
Apr. 30	15,958	44	12,996	8	5	2,910	35,690
May 31	18,336	73	12,956	7	22	5,278	40,722
June 30	23,864	70	18,609	17	36	5,132	45,384
July 31	20,162	224	11,767	9	10	8,152	52,712
Aug. 31	17,286	115	11,528	5	40	5,598	57,628
Sept. 30	16,175	199	13,484	5	83	2,404	59,556
Oct. 31	12,176	2	8,923	....	1	3,250	62,142
Nov. 30	13,284	2	9,129	....	2	4,151	65,489
Total	239,274	1,125	177,419	168	373	60,189	....

[1] Rejections by commanding officer include minors whose parents refused consent, married men whose wives refused consent, and men with criminal records or who were otherwise undesirable.

[2] Rejections by medical officer include all rejections at recruiting office as well as those rejected by the medical officer at the recruit depot to which they were transferred.

ENLISTMENTS BY STATES.

The following table shows the number of men enlisted in the Marine Corps, not including reserves enrolled but including inductees, between April 1, 1917, and November 11, 1918. These figures do not include the 13,214 enlisted men already in the Marine Corps on April 6, 1917:

Alabama	313	Nevada	86
Arizona	210	New Jersey	1,251
Arkansas	290	New Hampshire	67
California	2,527	New Mexico	25
Colorado	1,262	New York	6,782
Connecticut	240	North Carolina	488
Delaware	72	North Dakota	225
District of Columbia	451	Ohio	4,968
Florida	110	Oklahoma	384
Georgia	674	Oregon	1,006
Illinois	4,959	Pennsylvania	4,365
Idaho	508	Rhode Island	64
Indiana	1,182	South Carolina	66
Iowa	607	South Dakota	145
Kansas	673	Tennessee	1,418
Kentucky	592	Texas	2,205
Louisiana	832	Utah	898
Maine	24	Vermont	21
Massachusetts	1,957	Virginia	617
Maryland	867	Washington	1,767
Michigan	2,115	West Virginia	598
Minnesota	2,581	Wisconsin	876
Missouri	3,721	Wyoming	92
Mississippi	297		
Montana	1,205	Total	57,144
Nebraska	461		

Statistics that will show the exact number of officers and enlisted men from each State are being prepared.

Chapter IV.

GEOGRAPHICAL LOCATION AND DISPOSITION OF MARINES DURING WAR.

During the period of the war Marines served ashore and afloat all over the world. The following tables show where they were located at the outbreak of war and on the date the armistice became operative; also the naval vessels on which Marines were serving on both of these dates; and the geographical location of Marines during the war.

Location of Marines on April 6, 1917, and November 11, 1918.

Location.	Apr. 6, 1917.			Nov. 11, 1918.		
	Officers.	Men.	Total.	Officers.	Men.	Total.
American Expeditionary Forces				[1] 857	23,698	24,555
Azores				11	188	199
China	8	268	276	11	271	282
Cuba	16	580	596	99	2,310	2,409
England (A. E. F.). *See* American Expeditionary Forces.						
England (not A. E. F.)				2	69	71
France (A. E. F.). *See* American Expeditionary Forces.						
France (not A. E. F.)				146	1,030	1,176
Germany (A. E. F.) *See* American Expeditionary Forces.						
Guam	9	383	392	14	366	380
Haiti	62	622	684	60	825	885
Hawaiian Islands	3	137	140	10	466	476
Holland (The Hague)					3	3
Nicaragua	3	111	114	5	118	123
Philippine Islands	7	272	279	12	582	594
Porto Rico (San Juan)				1	77	78
Samoa				1		1
Santo Domingo	69	1,856	1,925	84	1,879	1,963
Sea duty	49	2,187	2,236	64	2,009	2,073
Spain (Madrid)					1	1
United States	183	6,481	6,664	1,029	36,004	37,043
Virgin Islands	10	317	327	25	583	608
Total	419	13,214	13,633	2,431	70,489	72,920

[1] Including enlisted men commissioned in Europe.

MARINES SERVING ON BOARD NAVAL VESSELS.

Marine detachments served on board all the overseas battleships and on the battleships of Battleship Force Two throughout the war. The Marines of Battleship Force One of which the *Minnesota* was flagship were temporarily withdrawn in April, 1918.

Marines were also on board a great many of the cruisers which acted as escorts for the vessels transporting Army troops to Europe.

The following table shows in detail those vessels which carried Marine detachments at the beginning of the war and on Armistice Day:

Ship.	Apr. 6, 1917.		Nov. 11, 1918.	
	Officers.	Men.	Officers.	Men.
Atlantic Fleet	1		1	
Pacific Fleet	1		1	
Asiatic Fleet	1		1	
Battleship Force 2			1	
Battleship Force	1			
Cruiser Force			1	
Division 6	1		1	
Division 7	1			
Division 8			1	
Division 9 (Sixth Battle Squadron)			1	
Alabama	1	40		
Arizona	2	83	2	88
Arkansas	1	76	2	86
Brooklyn	2	69	2	98
Castine		20		
Charleston			2	62
Cincinnati	1	40	1	41
Columbia		19		
Connecticut	3	65		
Constellation		6		7
Delaware	1	65	2	70
Denver	1	40		
Des Moines		38		
Dolphin		15		20
Florida	1	66	2	63
Frederick			2	64
Galveston	1	39	1	40
George Washington			2	97
Helena	1	30	1	25
Huntington			2	61
Idaho			2	19
Louisiana	1	64		
Machias		20		
Mayflower		15		5
Michigan	2	62		
Minnesota	2	68		
Mississippi			2	78
Montana	1	62	2	72
Nebraska	1	68		
Nevada	1	77	2	79
New Hampshire	1	67		
New Jersey	1	6		
New Mexico			2	68
New York	1	77	2	20
North Carolina			2	65
North Dakota	1	64	2	65
Oklahoma	2	77	2	80
Olympia	1	40		
Pennsylvania	1	94	3	133
Pittsburgh	2	75	2	105
Prairie		19		
Pueblo	1	69	2	15
Rhode Island	1	64		
Seattle	1	61		
South Carolina	2	65		
South Dakota			2	59
St. Louis			2	62
Texas	1	72	2	78
Utah	2	62	2	72
Wilmington	1	30	1	30
Wyoming	1	78	2	82
Yorktown		20		
Total	49	2,187	64	2,009

In addition to the above-named vessels, Marines served on the *Leviathan*, *Albany*, *New Orleans*, *Georgia*, *Kansas*, *Vermont*, *San Diego*, and *Virginia*.

GEOGRAPHICAL LOCATION OF MARINES DURING THE WAR.

During the period of the war Marines were stationed at the following posts:

UNITED STATES.

Navy yards and stations.—Portsmouth, N. H.; Boston; New York; Philadelphia; Annapolis; Washington, D. C.; Norfolk, Va.; Charleston, S. C.; Key West, Fla.; Pensacola, Fla.; New Orleans; Mare Island, Calif.; Puget Sound, Wash.; and North Island, Calif.

Naval magazines.—Hingham, Mass.; Fort Lafayette; Iona Island, N. Y.; Lake Denmark, N. J.; Fort Mifflin, Pa.; St. Juliens Creek, Va.; and Mare Island, Calif.

Naval ammunition depots.—Dover, N. J., and New London, Conn.

Torpedo stations.—Puget Sound, Wash.,and Newport, R. I.

Radio stations, etc.—Greenbury, Md.; Point Isabel, Tex.; Radio, Va.; Key West, Fla.; Chatham, Mass.; Portland, Me.; Rye Beach, Me.; Otter Cliffs, Me.; naval radio station, Wellfleet, Mass.; French Cable Co., Orleans, Mass.; Postal Telegraph and Cable Co., Rockport, Mass.; Commercial Telegraph & Cable Co., Boston; Marconi Wireless Co., Boston; Western Union Co., Boston; Cape Cod, Mass.; Sayville, N. Y.; New Brunswick, N. J.; Belmar, N. J.; Tuckerton, N. J.; Beaufort, S. C.; Charleston, S. C.; Annapolis, Md.; Washington, D. C.; San Diego, Calif.; Chollas Heights, Calif.; Point Arguello, Calif.; Inglewood, Calif.; East San Pedro, Calif.; Eureka, Calif.; Bolinas, Calif.; Marshall, Calif.; Farallones Islands, Calif.; Marshfield, Oreg.; Astoria, Oreg.; Lents, Oreg.; Tatoosh, Wash.; North Head, Wash.

Naval prisons.—Portsmouth, N. H.; Parris Island, S. C.; and Mare Island, Calif.

Naval hospitals.—Boston; New York; Washington, D. C.; Norfolk, Va.; Key West, Fla.; and Fort Lyons, Col.

Coaling stations.—La Playa, Calif., and Tiburon, Calif.

Receiving ship.—Boston.

Other places.—Headquarters, Washington, D. C.; Office of the Judge Advocate General; assistant paymasters' offices at New York, Atlanta, Ga., and San Francisco, Calif.; depots of supplies at Philadelphia, Pa., San Francisco, Calif., and Charleston, S. C.; naval experimental station, New London, Conn.; naval district base, New London, Conn.; advanced base force, Philadelphia, Pa.; mobilization bureau, New York City; third naval district base, New York; New Navy Building guard, Washington, D. C.; naval mine station, Yorktown, Va.; naval base, Hampton Roads, Va.; Navy rifle range, Wakefield, Mass.; rifle range, Winthrop, Md.; naval proving grounds, Indian Head, Md.; Wissahickon Barracks, N. J.; Navy fuel depot, Curtis, Md.; Navy ordnance plant, Charleston, W. Va.; camp of instruction, bayonet team, Lansdowne, Pa.; signal battalion, Paoli, Pa.; staff office, San Francisco, Calif.; Marine barracks, Quantico, Va.; Fort Crockett, Galveston, Tex.; Gerstner Field, Lake Charles, La.; naval air station, Cape May, N. J.; naval air station, San Diego, Calif.; naval school for mechanics, Great Lakes, Ill.; naval air station, Pensacola, Fla.; Army training field, Mineola, Long Island, N. Y.; Marine Corps School of Machine Gun Instruction at Utica, N. Y.; and Massachusetts Institute of Technology, Boston, Mass.

BEYOND CONTINENTAL LIMITS OF UNITED STATES.

American Expeditionary Forces.—In France, England, and Germany.

With naval service in Europe.—Paris, France; Pauillac, France; London, England; Marine aerodromes between Calais and Dunkirk, France; Croix d'Hins, Gironde, France; naval base, Ponta Delgada, Azores Islands; Cardiff, Wales.

Naval stations.—Cavite, P. I.; Olongapo, P. I.; Pearl Harbor, Hawaii; Guantanamo Bay, Cuba; Virgin Islands; Guam.

Occupation forces.—Santo Domingo, Haiti.

Legation guards.—Peking, China; and Managua, Nicaragua.

Couriers.—Madrid, Spain; The Hague, Holland; Luxembourg; Jassy, Roumania; Stockholm, Sweden; Copenhagen, Denmark; Christiania, Norway; Petrograd, Archangel, Murman Coast, Russia; Paris, France; London, England; Athens, Greece; and Rome, Italy.

Constabularies.—Guardia Nacional Dominicana and Haitian gendarmerie.

Radio stations.—Cavite, P. I.; San Juan, Porto Rico; El Cayay, Porto Rico; Haiti; Croix d'Hins, Gironde, France.

Naval ammunition depot.—Olongapo, P. I.

Naval magazine.—Pearl Harbor, Hawaii.

Depot of supplies.—Cavite, P. I.

Attachés.—Paris, France; London, England; Yokohama, Japan; Petrograd, Russia; Stockholm, Sweden; Copenhagen, Denmark; and Christiania, Norway.

Chapter V.

HOW OFFICERS WERE OBTAINED AND TRAINED.

HOW OFFICERS WERE OBTAINED.

The outbreak of war made it essential that the corps should be filled as far as practicable with officers who had had prior military experience and training, and immediate steps were taken to arrange for the designation and examination of Marine Corps warrant officers, noncommissioned officers, graduates of military colleges, and other civilians with military experience and training.

The appointment of officers subsequent to the declaration of war up to October, 1917, both for the permanent service and for the temporary increase authorized for the duration of the war, were drawn from the following sources:

Graduates of the Naval Academy	6
Former officer of the Marine Corps	1
Former graduate of the Naval Academy	1
Warrant officers and paymaster's clerks of the Marine Corps	89
Meritorious noncommissioned officers of the Marine Corps	122
Reserve officers and National Naval Volunteers	36
Graduates of military colleges	284
Other civilians with prior military or naval experience or training	136
Other civilians passing the competitive examination held July 10, 1917	86

In order to expedite the training of the new officers, advantage was taken of the law providing for a Marine Corps Reserve, and successful candidates were immediately enrolled as second lieutenants in the reserve and ordered to Marine Corps posts for instruction pending the issuance of their commissions in the regular service. Candidates designated for the examination held July 10, 1917, were authorized upon designation to enroll as privates in the Marine Corps Reserve, with the understanding that upon the completion of their examination they would be ordered to the Recruit Depot at Parris Island, S. C., for training pending the receipt of the report of the examining board. This policy was carried out, and the successful candidates were commissioned second lieutenants in the Marine Corps, while the unsuccessful candidates were given the option of continuing in the service as enlisted men or of being discharged therefrom.

Owing to the unusually large number of young men of excellent education and fine attainments who had enlisted in the Marine Corps after the outbreak of war, it was decided that no further appointments of civilians to the rank of second lieutenant would be made during the continuance of the war, and that all vacancies occuring in that grade, not required for graduates of the Naval Academy, would be filled by the promotion of meritorious noncommissioned officers This decision was promulgated to the service in Marine Corps Orders No. 25 (Series 1917).

HOW OFFICERS WERE TRAINED.

The officers appointed from civil life, as soon as enrolled, were ordered to the Marine barracks, Mare Island, Calif.; San Diego, Calif.; Parris Island, S. C.; and the Marine Corps rifle range, Winthrop, Md., for instruction, pending the completion of the buildings for their use at the Marine barracks, Quantico, Va. Early in July, 1917, the buildings being in readiness, the newly appointed officers, about 345 in number, were assembled at Quantico, where an officers' camp of instruction was held, and the course completed in October, 1917.

In carrying out the policy of obtaining officers from the ranks, orders were issued to commanding officers of every post and station of the Marine Corps, both at home and abroad, as well as those on board ship, to the effect that all commissioned officers would be taken from the ranks, and that the number of men to be designated from each post to attend the training camp would be a certain percentage of the number of men stationed at such post or station. Each commanding officer was ordered to convene a board of three officers to examine into the qualifications of the men at his post, and to report in the order of merit the names of the men considered qualified for entry to the officers' training camp at Quantico, Va. These reports were all forwarded to headquarters, where a board was convened to examine them and to select, in accordance with their standing as reported by the various boards, the number of men who it had been decided could be quartered and properly instructed at Quantico. It was found that about 600 was the limit that could be accommodated, and approximately this number was selected for the first camp, which was established at Quantico, Va., in April, 1918.

The officers' training camp was commanded by an officer of adequate rank. The students were divided into companies with a major in command as chief instructor and captains and lieutenants to assist him. The candidates were given a very rigid course of instruction and intensive training. Some of the studies pursued were: Infantry drill regulations, manual of interior guard duty, bayonet training, bombing, minor tactics, military engineering, military topography, administration, military law, lectures on gas and on sea duty, and a practical course on the rifle range.

The training at these camps was most intensive and thoroughly competitive, so that a man's position depended entirely upon himself. The material to draw from was so excellent that comparatively few of those who entered the camps failed to receive commissions and many of the young men so commissioned who were assigned to duty abroad demonstrated that their selection was fully justified.

Many officers also received special training in the schools of the Overseas Depot at Quantico, Va.

The majority of the members of the first officers' training camp were graduated in July, 1918. Three hundred of this camp were commissioned on July 15, 1918, and 91 on August 15, 1918.

The same proportionate allowance that was made in the United States was also designated for the Marines serving in France, and similar means were instituted there to carry out the policy of selection of men for the training camp. As a result of the camp established over there, 164 second lieutenants were appointed from the Fourth Brigade in France.

The second officers' training camp was opened at Quantico, Va., on August 20, 1918, the enlisted men forming its personnel having been selected in exactly the same manner as those attending the first camp and this procedure was also followed with regard to the Marines of the American Expeditionary Forces in France. Of the 570 men enrolled 432 were graduated from the second officers' training camp, December 16, 1918, and 172 from the Army candidate school in France, who, immediately upon graduating, were enrolled as second lieutenants (provisional) in class 4, Marine Corps Reserve, and subsequently appointed temporary second lieutenants in the Marine Corps. An extension of three weeks to this course in America was necessitated by the epidemic of influenza.

There were 235 graduated in July, 1919, from the third officers' training camp who were enrolled as second lieutenants (provisional) in class 4, Marine Corps Reserve, and immediately assigned to inactive duty.

There were also 48 graduates of the Army candidate school in France, who were enrolled as second lieutenants (provisional) in class 4, Marine Corps Reserve, and who were discharged or placed on inactive duty upon their return to the United States, with the exception of four who were transferred to the temporary service.

Sixty-nine officers were graduated from the Marine Corps school of machine-gun instruction at Utica N. Y.

Information regarding the training of Marine officers for aviation duties will be found in Chapter XXI.

MARINE SECTIONS, STUDENT ARMY TRAINING CORPS.

In the act approved August 31, 1918, provision was made for a Student Army Training Corps, and under date of September 12, 1918, the Secretary of War directed the Provost Marshal General to allot 1,500 of the registrants authorized for induction into the Student Army Training Corps to the Marine sections under that organization. On September 23, 1918, with the approval of the Navy Department, Marine Corps headquarters designated the following institutions for the organization of Marine sections of the Student Army Training Corps and allotted quotas to each ranging from 100 to 190:

Leland Stanford Junior University	110
Georgia School of Technology	100
Harvard University	120
University of Minnesota	110
Cornell University	170
University of Washington	160
University of Texas	100
Yale University	100
University of Kansas	140
University of Wisconsin	190
Virginia Military Institute	100
University of North Carolina	100

A Marine officer was ordered to each of the designated institutions and charged with the duties of administration, instruction, and discipline of the Marine section, with the assistance of a noncommissioned officer of the Marine Corps.

It was intended to transfer, from time to time, well-qualified students who were inducted into Marine sections of the Student Army Training Corps to aviation duty, or to one of the two recruit camps, and in both cases men thus recommended, who proved themselves qualified to become officers, would be ultimately commissioned in either the Marine Corps Reserve Flying Corps or for general service in the Marine Corps. In either case after finishing their course in the Student Army Training Corps they would have been sent to a recruit camp for the regular course of training, because this would make it possible to imbue them with the necessary esprit de corps and indoctrinate them with the Marine Corps methods of procedure and training, both essential to the making of a Marine officer of the highest type. Owing to the ending of active hostilities there were no graduates from the Marine sections of the Student Army Training Corps at the different universities and colleges as they were ordered abandoned shortly after the armistice became operative

Chapter VI.

TRAINING OF ENLISTED MEN IN THE UNITED STATES AND EUROPE.

IN UNITED STATES.

The Marine Corps system of training for the enlisted personnel during the war was thorough and excellent in every respect, and resulted in the turning out of men who proved themselves well fitted for the arduous duties of Marines.

For a short time after the outbreak of the war temporary recruit depots were opened at the navy yards at Philadelphia, Pa., and Norfolk, Va., with a capacity of 2,500 at the former and 500 at the latter. These were used until the regular recruit depots at Parris Island, S. C., and Mare Island, Calif., could accommodate the recruits. These two recruit depots were greatly enlarged both in size and scope, to take care of the temporary increase in strength authorized for the war, and were soon able to meet all demands made upon them.

At the beginning of the war the course of recruit instruction at the recruit depot, Parris Island, was of 8 weeks duration, and with but very few exceptions every recruit passing through this depot received 8 weeks instruction. At the Mare Island recruit depot, the recruits received 12 weeks training from April 6 to 28, 1917, 9 weeks from April 29, 1917, to June 21, 1918, and 8 weeks from June 22 to November 11, 1918.

The following table gives a list of the special schools at the Parris Island recruit depot and the number of graduates from each during the period between the outbreak of war and the date the armistice became operative:

Noncommissioned Officers School	2,144
Field Musics School	493
Radio School	143
Signal School	232
Band School	247
Clerical School	236
Pay School	78
Cooks, and Bakers, School	150
Total	3,723

The following table illustrates what was accomplished by the two recruit depots:

Depot.	In training—		Maximum strength of post.	Maximum number of recruits at one time.	Total recruits handled.	Maximum capacity.
	Apr. 6, 1917.	Nov. 11, 1918.				
Parris Island	835	4,104	16,601	13,286	46,202	13,060
Mare Island	358	1,143	2,799	2,470	11,901	3,000
Total	1,193	5,247	19,400	15,756	58,103	16,060

After leaving the recruit depots at Parris Island and Mare Island, advanced training was given the men at Quantico, Va. This training was most intensive and as a result all the organizations which were trained there attained a high state of efficiency. It was made to approximate as nearly as practicable the real service which the men would have in the American Expeditionary Forces in France. Officers who were engaged in this training showed great ingenuity and efficiency in their attempts to make the training approach as nearly as possible what the men would be subiected to in actual service. That they succeeded was shown by the work done by the Marines in France and other places.

The first troops arrived at the Marine barracks, Quantico, Va., on May 18, 1917. The maximum enlisted strength was 9,849 on September 12, 1918. The maximum number of officers present at one time was 484, on August 16, 1918. The strength on November 11, 1918, was 329 officers and 8,798 enlisted men. From May, 1917, to November 11, 1918, approximately 1,000 officers and 40,000 enlisted men passed through Quantico, Va.

In addition to giving the enlisted men general training at Quantico in preparation for overseas and other duty, the Overseas Depot was established on May 19, 1918, for the double purpose of organizing and training units of the Marine Corps for service with the American Expeditionary Forces.

Prior to the organization of this depot the Fifth and Sixth Regiments, the Sixth Machine Gun Battalion, the Base Battalion of the Fifth Regiment, and two replacement battalions had left the United States and had become part of the American Expeditionary Force.

The Overseas Depot consisted of an administrative staff and the various sections as follows: (*a*) The specialists' schools for the technical training of the infantry and machine gun, and the coordination of these specialists' arms; (*b*) the tactical department for the instruction and training of overseas units in new tactical principles; (*c*) the enlisted staff school for the training of first sergeants, mess sergeants, cooks, company clerks, armorers, etc. Two French and four Canadian officers, who had abundant experience in the fighting in Europe, were assigned as advisors of the commanding officer.

The basic independent unit of organization was the platoon, and the platoon therefore became the principal training unit. In the organization of this unit the scheme followed was to assure to each a certain nucleus of enlisted instructors trained in the various specialties, in addition to the platoon commanders, who were qualified to carry on the instruction along approved lines within the unit. This nucleus was taken from the graduates of the specialists' schools of the depot. When four such platoons had been formed they were assembled into a company. The company headquarters, trained in the enlisted staff school, was added to the four platoons and the company organization was turned over to the company commander complete in all details. Battalions were likewise formed by the consolidation of companies. In every instance the platoon, company, and the battalion, carried out a regular schedule of drills and instructions under the supervision of the depot, but all administrative details were left in the hands of the company and the battalion commanders. These training schedules were made up in the tactical department, approved by the commanding officer, and were based on the most approved methods in effect at the time. In the cases of the formation of regimental organizations, of which there were two formed during the existence of the Overseas Depot, the battalions upon being formed were turned over to the regimental commander, and in this case direct supervision by the depot ceased, but all facilities on hand, such as material, officers acting in an advisory capacity, training areas, etc., directly attached to the depot, were placed at the disposal of the regimental commanders who were at all times in active liaison with the depot.

About 85 per cent of the troops forming the detachments arriving at the Overseas Depot for service in France had undergone not less than 8 nor more than 12 weeks' training at the regular recruit depots of the Marine Corps. The preliminary training received at these recruit depots was such as to fit the men for general service throughout the Marine Corps, and resulted in the men being well disciplined, considering the short time they had been in the service. This facilitated the more advanced and specialized training they were to receive at the Overseas Depot. These detachments were composed entirely of qualified riflemen, having undergone during the recruit period a most thorough and comprehensive course in the use of the rifle. Upon the arrival of these detachments they were organized as outlined above, and the commissioned personnel was assigned to the units from the officers' school. The schedule and drills and instructions were provided them and were carried out under the supervision of specially selected officers of the tactical department of the Overseas Depot, including the foreign officers. This training continued until the units departed for France. Training in open warfare was given precedence over that of trench warfare from the very beginning in the proportion of about four to one.

The following units were organized by the Overseas Depot: Third, Fourth, Fifth, Sixth, Ninth, Tenth, Eleventh Separate Battalions; Second and Third Machine Gun Battalions; Fifth Brigade Machine Gun Battalion; Second and Third Separate Machine Gun Battalions; Eleventh and Thirteenth Regiments; total, approximately, 16,000 officers and enlisted men. The Seventh and Eighth Separate Battalions were organized and sent to France from Marine Barracks, Parris Island, S. C.

The following table shows the schools conducted by the Overseas Depot and the number of graduates:

School.	Officers.	Noncommissioned officers.	Privates.	Total.
Officers	294			294
Bayonet		360	220	580
Bombing		200	150	350
Gas		180		180
Automatic Rifle		150	650	800
Scout Snipers		75	375	450
Machine Gun	120	145	295	560
Miners and Sappers		70	80	150
Enlisted Staff		105	300	405
Total	414	1,285	2,070	3,769

In addition to the training described above, 69 officers and 2,084 enlisted men, a total of 2,153, graduated from the Marine Corps school of machine-gun instruction at Utica, N. Y.

Never before in the history of the corps have better drilled and trained or more generally efficient men been turned out, ready for duty, upon completion of their training, and to this factor is largely due the splendid record made by the Marines during the war.

The work of the officers training the Marines was not spectacular, and they wear no war chevrons, nor decorations for bravery, perhaps, but they were, nevertheless, a vital factor in whatever success the Marine Corps met with in the great struggle.

Information with reference to the training of enlisted men for aviation will be found in Chapter XXI.

TRAINING IN FRANCE.

On June 27, 1917, the First Battalion of the Fifth Regiment actually landed in France and on July 3, 1917, the entire Fifth Regiment was under canvas on French soil. From that date every effort was made to train the men and officers. Elements of the Fifth Regiment trained as a part of the First Division of Regulars from July 15, 1917, to September, 1917, in the Gondrecourt training area. From September, 1917, on, the training of the available units of the Fourth Brigade as a unit of the Second Division of Regulars was conducted in the Bourmont training area.

Until February, 1918, the training of the Marines in France was handicapped by the fact that units of the Brigade were engaged in duties along the Line of Communications (Services of Supply), one company and a battalion commander being absent in England until March, 1918. It was not until the middle of February, 1918, that the Fourth Brigade of Marines (less the company in England) was conducting its training as a brigade with any degree of satisfaction. Owing to the well-trained condition of the individual Marine this condition did not vitally affect his professional ability as was so distinctly shown by his later accomplishments.

The Fourth Brigade continued its training in the Bourmont training area until the middle of March, 1918, when it entered the front line trenches in the Verdun sector.

The Marine replacements received little or no training in a training area in France as most of them were hurried into the fighting immediately upon arrival overseas.

To summarize, the average Marine who arrived in France received at least six weeks' training in the United States in a recruit depot and a very short period at Quantico. This is a contrast to the six months' training received by the average enlisted man of the Army. After arrival in France the Marines, except those of the original Fourth Brigade, received practically no training in a training area since they joined the brigade almost immediately. The Marines comprising the Fifth Brigade of Marines received no training in a regular training area in France.

Chapter VII.

ORGANIZATIONS AND REPLACEMENTS SENT TO EUROPE—ORGANIZATION OF THE FOURTH AND FIFTH BRIGADES.

THE FOURTH BRIGADE OF MARINES.

The Fourth Brigade of United States Marines was composed of the Fifth and Sixth Regiments of Marines, and the Sixth Machine Gun Battalion of Marines.

The companies forming the battalions were as follows:

FIFTH REGIMENT.

First Battalion.	*Second Battalion.*	*Third Battalion.*
17th (A) Company.	18th (E) Company.	16th (I) Company.
49th (B) Company.	43d (F) Company.	20th (K) Company.
66th (C) Company.	51st (G) Company.	45th (L) Company.
67th (D) Company.	55th (H) Company.	47th (M) Company.

8th Machine Gun Company.
Supply Company.
Headquarters Company.

SIXTH REGIMENT.

First Battalion.	*Second Battalion.*	*Third Battalion.*
74th (A) Company.	78th (E) Company.	82nd (I) Company.
75th (B) Company.	79th (F) Company.	83rd (K) Company.
76th (C) Company.	80th (G) Company.	84th (L) Company.
95th (D) Company.	96th (H) Company.	97th (M) Company.

73d Machine Gun Company.
Supply Company.
Headquarters Company.

SIXTH MACHINE GUN BATTALION.

15th (A) Company.	77th (C) Company.
23d (B) Company.	81st (D) Company.

From June 27, 1917, to the middle of September, 1917, the Fifth Regiment was a unit of the First Division of Regulars. Although the Fifth Regiment was the only organization of Marines in France at the time, the Fourth Brigade of Marines was formed on October 23, 1917, when Col. Charles A. Doyen cabled acceptance of his appointment as Brigadier General. From October 26, 1917, to August 8, 1919, the Fourth Brigade was a part of the Second Division of Regulars, except from October 20–23, 1918, when the Brigade was provisionally at the disposal of the Ninth French Army Corps, in the vicinity of Leffincourt. On August 8, 1919, the brigade was transferred back to the naval service.

On May 29, 1917, in accordance with directions issued by the President, the Secretary of the Navy directed the Major General

Commandant "to organize a force of Marines to be known as the Fifth Regiment of Marines for service with the Army as a part of the first expedition to proceed to France in the near future." The Fifth Regiment was accordingly organized at the navy yard, Philadelphia, Pa., on June 7, 1917, with Col. Charles A. Doyen in command, and Maj. Harry R. Lay, as adjutant.

Gen. Pershing and his staff, accompanied by two Marine officers, preceded the first expedition to France, sailing late in May, 1917, from the United States.

The final report of the American commander in chief includes the following:

> The offer by the Navy Department of one regiment of Marines to be reorganized as Infantry was accepted by the Secretary of War, and it became temporarily a part of the First Division.

On June 14, 1917, the first expedition of American troops left the United States for France and the Fifth Regiment of Marines embarked on the naval transports *Henderson* and *Hancock,* and the auxiliary cruiser *De Kalb* (former *Printz Eitel Friedrich*), formed approximately one-fifth of it. The fourth group, including the *Hancock,* did not sail until June 17, 1917.

The orders received by the convoy commander on the day prior to sailing read in part: "A military expedition is to be embarked on the above-named transports, augmented by a regiment of Marines embarked in naval vessels, for transportation to a destination already communicated."

The *De Kalb* was in group 1, the *Henderson* in group 2, and the *Hancock* in group 4; all were part of the escort and not the convoy.

Rear-Admiral Albert Gleaves, the convoy commander, flying his flag on the *Seattle,* personally commanded the first group, while Maj. Gen. W. L. Sibert in the *Tenedores,* was the senior Army officer embarked.

The passage of the four groups across the Atlantic was successfully accomplished without a single disaster, or the loss of a life due to enemy causes.

At 10.15 p. m., June 22, 1917, the first group, including the *De Kalb,* was attacked by enemy submarines. The wake of a submarine was sighted crossing 50 yards ahead of the *Seattle's* bow from starboard to port. A few seconds later the *De Kalb* and *Havana* sighted torpedoes and opened fire. Two torpedoes passed close to the *Havana,* and one passed ahead and one astern of the *De Kalb.* The second group encountered two submarines, the first at 11.50 a. m., June 26, 1918, about 100 miles off the French coast and the second two hours later.

The *De Kalb* arrived at St. Nazaire, France, on June 26, 1917, the *Henderson* on June 27, 1917, and the *Hancock* on July 2, 1917. On June 27, 1917, the commanding officer of the Fifth Regiment reported to the commanding general, First Division, American Expeditionary Forces, and from that date the Fifth Regiment was considered as being detached for service with the Army by direction of the President.

Five hundred negro stevedores had been brought from the United States by the Army to discharge ships, but they were found inadequate for the large number of ships concerned. The Marines relieved the situation somewhat by turning to and discharging their own vessels.

On June 27, 1917, the First Battalion, less the Fifteenth Company which joined the battalion the following day, disembarked from the *De Kalb* and occupied quarters ashore. On this date Lieut. Col. Logan Feland joined the Fifth Regiment. On June 28, 1917, the Second and Third Battalions went ashore from the *Henderson* for a practice march, and the following day the First Battalion erected tents for the regiment on a camp site a short distance outside of St. Nazaire. By 8 p. m., July 3, 1917, the entire Fifth Regiment was ashore under canvas.

On July 15, 1917, the Fifth Regiment, less the Third Battalion, which remained behind to perform guard duty and other detached units and officers, proceeded to the Gondrecourt training area, and was stationed in Menaucourt and Naix.

On August 1, 1917, Gen. Pershing inspected the battalions at the two towns where they were billetted.

On August 15, 1917, the First Division, including the Fifth Regiment of Marines, was reviewed by its commanding general on a plateau 12 miles distant from the training area.

On August 19, 1917, Gen. Pershing and Gen. Petain, commander-in-chief of all the French forces, inspected the Marines, as a unit of the First Division. Gen. Petain congratulated the colonel of the regiment on the splendid appearance of its officers and men, as well as the cleanliness of the towns.

Every opportunity was taken advantage of to perfect the regiment for combat duty, but this work was handicapped by the fact that many units of the regiment were scattered along the Line of Communications performing duty of a necessary but of a nontraining nature. One company and one battalion commander left the regiment on September 22, 1917, for duty in England, and did not rejoin the regiment until March 11, 1918. Many other officers and men were placed on detached duty.

On September 24, 25, 1917, that part of the Fifth Regiment available for training arrived in the Bourmont training area and was stationed at Damblain and Breuvannes.

The following letter dated November 10, 1917, addressed by Gen. Pershing to the Major General Commandant is both complimentary and explanatory as to why the Marines were used along the Line of Communications:

> Your Marines having been under my command for nearly six months, I feel that I can give you a discriminating report as to their excellent standing with their brothers of the Army and their general good conduct. I take this opportunity, also, of giving you the reasons for distributing them along our Line of Communications which, besides being a compliment to their high state of discipline and excellent soldierly appearance, was the natural thing to do as the Marine Regiment was an additional one in the Division and not provided for in the way of transportation and fighting equipment in case the Division should be pushed to the front. When, therefore, service of the rear troops and military and provost guards were needed at our base ports and in Paris it was the Marine Regiment that had to be scattered, in an endeavor to keep the rest of the organized division intact.
>
> I have been obliged to detach a number of your officers as assistant provost marshals in France and in England, all of which I take it you will agree with me was highly complimentary to both officers and men, and was so intended. I can assure you that as soon as our service of the rear troops arrive, including a large number of officers and men for the specific duties now being performed by your men, the Marines will be brought back once more under your brigade commander and assigned to the duties which they so much desire in the Second Regular Division under General Bundy.
>
> It is a great pleasure to report on your fine representatives here in France.

151402°—20——3

Col. Charles A. Doyen was in command of the Fifth Regiment from the date of its organization on June 7, 1917, to October 29, 1917; and Lieut. Col. Hiram I. Bearss from October 30, 1917, to December 31, 1917. Col. Wendell C. Neville having arrived on on board the *De Kalb* at St. Nazaire, France, on December 28, 1917, reported to the Fourth Brigade for duty on January 1, 1918, and on that date assumed command of the Fifth Regiment, continuing in command until July, 1918.

The Sixth Machine Gun Battalion of Marines was organized at the Marine barracks, Quantico, Va., by order of the Major General Commandant on August 17, 1917. The battalion was designated the First Machine Gun Battalion, but on January 20, 1918, after arrival in France, was renamed the Sixth Machine Gun Battalion. On December 14, 1917, the battalion sailed from New York on the *De Kalb*, arriving at St. Nazaire, France, December 28, 1917. On January 3, 1918, the battalion arrived at Damblain in the Bourmont training area and began training with headquarters at Germainvilliers.

Maj. Edward B. Cole was in command of the Sixth Machine Gun Battalion of Marines from the date of its organization until June 10, 1918, when he received a mortal wound.

On August 4, 1917, in accordance with directions issued by the President, the Secretary of the Navy directed the Major General Commandant "to organize a force of Marines, to be known as the Sixth Regiment of Marines, for service with the Army in France," and the regiment was organized as directed.

On September 23, 1917, the First Battalion of the Sixth Regiment sailed on the *Henderson* from New York and landed at St. Nazaire, France. on October 5, 1917. On October 17, 1917, the Seventy-third Machine Gun Company, Headquarters, and Supply Companies, and Col. Albertus W. Catlin, commanding officer of the Sixth Regiment, with his Staff, sailed from Philadelphia, Pa., on the *De Kalb*, and from New York on October 18, 1917, arriving at St. Nazaire, France, on November 1, 1917. On October 31 1917, the Third Battalion of the Sixth Regiment sailed from New York on board the *Von Steuben* and anchored at Brest, France, on November 12, 1917. On January 24, 1918, the Second Battalion of the Sixth Regiment sailed on the *Henderson* from New York and arrived at St. Nazaire, France, February 6, 1918, and with the arrival of this last battalion, the entire Sixth Regiment of Marines was in France.

On October 23, 1917, the Fourth Brigade of Marines was organized, with Brig. Gen. Charles A. Doyen in command. Brig. Gen. Doyen continued in command until May 7, 1918, when he published in General Orders No. 5, that he had relinquished command. Maj. Harry R. Lay was the first brigade adjutant, and performed the duties of that office from October 24, 1917, to August 9, 1918, except during the period February 7 to May 9, 1918, when Maj. Holland M. Smith was brigade adjutant.

On October 26, 1917, Brig. Gen. Charles A. Doyen, United States Marine Corps, assumed command of the Second Division as its first commanding general, and announced his staff in General Orders No. 1, with station at Bourmont, Haute-Marne, serving as such until relieved by Maj. Gen. Omar Bundy, United States Army, who announced that he assumed command in General Orders No. 4, November 8, 1917.

Like the Fifth Regiment, the Sixth Regiment spent several months performing the necessary but undesired duties along the Line of Communications. On January 12, 1918, Col. Albertus W. Catlin established headquarters for the Sixth Regiment at Blevaincourt in the Bourmont training area. The Third Battalion arrived in this area on January 12, 1918, the headquarters units the same date, the First Battalion during January, 1918, and the Second Battalion on February 10, 1918.

Therefore, on February 10, 1918, the Fourth Brigade of Marines was in the Bourmont training area intact, with the exception of one company on duty in England, training industriously as an infantry brigade of the Second Division. While the brigade had been organized on October 23, 1917, and had actually functioned as a brigade with elements of all three of its units present from January 12, 1918, it was not until February 10, 1918, that the Brigade organization was perfected.

FIFTH BRIGADE OF MARINES.

On September 5, 1918, the Major General Commandant directed the post commander, Marine barracks, Quantico, Va., to organize brigade headquarters of the Fifth Brigade, United States Marine Corps.

This brigade was accordingly organized and was composed of the Eleventh and Thirteenth Regiments and the Fifth Brigade Machine Gun Battalion. The companies of the Fifth Brigade were designated by letters and not by numbers.

Brig. Gen. Eli K. Cole was designated as the brigade commander and on September 15, 1918, he and the Brigade Staff sailed from Hoboken, N. J., on board the *Von Steuben*, arriving at Brest, France, on September 24, 1918.

The Thirteenth Regiment left the Overseas Depot at Quantico, Va., on Friday, September 13, 1918, and on September 15, 1918, sailed from Hoboken, N. J., on board the *Henderson* and *Von Steuben*, arriving at Brest, France, on September 25, 1918.

On September 29, 1918, Eleventh Regiment Headquarters and the First Battalion sailed on the *De Kalb* from Philadelphia, Pa., and arrived at Brest, France, on October 13, 1918. On October 16, 1918, the Second and Third Battalions of the Eleventh Regiment sailed from Brooklyn, N. Y., on board the *Agamemnon* and *Von Steuben* and arrived at Brest, France on October 25, 1918.

On October 28, 1918, the Fifth Brigade Machine Gun Battalion sailed from South Brooklyn, N. Y., on board the *Henderson* and arrived at Brest, France, on November 9, 1918. With the arrival of this unit the entire Fifth Brigade was in France.

AVIATION UNITS.

On January 21, 1918, the First Marine Aeronautic Company arrived at naval base No. 13, Ponta Delgada, Azores.

On July 30, 1918, the First Marine Aviation Force (less Squadron D) disembarked at Brest, France, and formed the Day Wing of the Northern Bombing Group. Squadron D joined the Day Wing in October, 1918.

MARINE DETACHMENTS FOR NAVAL BASES.

On January 21, 1918, and on July 20, 1918, detachments for the naval base No. 13, arrived at Ponta Delgada, Azores.

On September 30, 1918, the detachment for naval base No. 29, arrived at Cardiff, Wales.

On December 29, 1918, the detachment for the naval forces in France, staff office, Paris, France, landed at St. Nazaire, France.

REPLACEMENTS FOR AMERICAN EXPEDITIONARY FORCES.

The following table will show the names of the replacement organizations sent to the American Expeditionary Forces, dates of sailing and arrival, and names of vessels:

Name of organization.	Date embarked in U. S.	Date disembarked in France.	Name of vessel.
Fifth Regiment Base Detachment	July 31, 1917	Aug. 22, 1917	Henderson.
Twelfth and Twenty-sixth Companies (disbanded in France).	Dec. 8, 1917	Dec. 31, 1917	De Kalb.
First Replacement Battalion	Feb. 5, 1918	Feb. 25, 1918	Von Steuben.
Second Replacement Battalion	Mar. 14, 1918	Mar. 27, 1918	Henderson.
Third Replacement Battalion	Apr. 22, 1918	May —, 1918	Do.
Casual Company	do	do	Do.
First Machine Gun Replacement Battalion	May 26, 1918	June 8, 1918	Do.
First Casual Replacement Battalion	do	do	Do.
Second Casual Replacement Battalion	June 30, 1918	July 9, 1918	Do.
Third Separate Battalion	Aug. 13, 1918	Aug. 26, 1918	Do.
Fourth Separate Battalion	do	do	Do.
Fifth Separate Battalion	Aug. 17, 1918	Aug. 27, 1918	Von Steuben.
Sixth Separate Battalion	do	do	Do.
First Separate Machine Gun Battalion	Aug. 21, 1918	Sept. 2, 1918	De Kalb.
Seventh Separate Battalion	Oct. 20, 1918	Nov. 3, 1918	Pocohontas.
Eighth Separate Battalion	do	do	Do.
Ninth Separate Battalion	Oct. 27, 1918	Nov. 9, 1918	Henderson.

In addition to the above the Twelfth Replacement Battalion sailed from the United States on board the *Hancock* in June, 1919, arrived in France in June, 1919, and joined the American Expeditionary Forces.

NUMBER OF MARINES SAILING FROM THE UNITED STATES TO EUROPE FOR DUTY WITH THE AMERICAN EXPEDITIONARY FORCES AND FOR SHORE DUTY WITH THE NAVAL SERVICE.

There were 834 officers, not including observers, and 30,481 enlisted men, or a total of 31,315 Marines, sent overseas for shore duty with the American Expeditionary Forces and naval service. The following tables give details:

For duty with American Expeditionary Forces.

Month of departure from United States.	Officers.	Enlisted men.	Total.
May, 1917	[1] 2	....	2
June, 1917	70	2,689	2,759
July, 1917	29	1,054	1,083
September, 1917	27	1,045	1,072
October, 1917	45	1,536	1,581
December, 1917	23	637	660
January, 1918	31	1,031	1,062
February, 1918	24	1,041	1,065
March, 1918	23	1,034	1,057
April, 1918	22	1,284	1,306
May, 1918	24	1,565	1,589
June, 1918	6	751	757
August, 1918	32	4,362	4,394
September, 1918	172	5,275	5,447
October, 1918	132	5,809	5,941
Total	662	29,113	29,775

[1] Accompanied Gen. Pershing.

Sixty officers of the Medical Corps, twelve officers of the Dental Corps, five hundred enlisted men of the Medical Corps, and eleven Chaplains, of the Navy, not included in the above figures, were sent to France and served with the Marines in the American Expeditionary Forces.

In addition to the above the Twelfth Replacement Battalion, consisting of 9 officers and 500 enlisted men, joined the American Expeditionary Forces in June, 1919.

For duty with naval service ashore.

Month of departure from United States.	Officers.	Enlisted men.	Total.
December, 1917	2	59	61
January, 1918	13	172	185
June, 1918	2	75	77
July, 1918	107	654	761
August, 1918	4	120	124
September, 1918	44	288	332
Total	172	1,368	1,540

Chapter VIII.

OPERATIONS IN GENERAL.

While the battle operations of the Fourth Brigade as an infantry brigade of the Second Division of Regulars overshadowed all others taken part in by Marine Corps personnel, those operations were by no means the only ones participated in by officers and men of the Marine Corps.

The commanding general of the Second Division from early in August, 1918, to the date of demobilization, and several officers on his staff were Marine officers. Officers of the Marine Corps were at various times attached to the First, Second, Third, Fourth, Sixth, Twenty-sixth, Thirty-second, Thirty-fifth, Ninetieth, and Ninety-second Divisions, and in some cases engaged in operations with them. Brig. Gen. John A. Lejeune assumed command of the Sixty-fourth Infantry Brigade of the Thirty-second Division, then in the front line on the Swiss border in the Suarce sector, on July 5, 1918. He was in command of this brigade on July 22, 1918, when it was withdrawn from the above-mentioned sector and continued in command until July 25, 1918, when he left to command the Fourth Brigade of Marines. Between July 5, 1918, and July 22, 1918, Brig. Gen. Lejeune, in addition to the Sixty-fourth Brigade, commanded three French infantry regiments. Col. Robert H. Dunlap was in command of the Seventeenth Field Artillery Regiment of the Second Field Artillery Brigade, Second Division, from October 30, 1918, to February, 1919. Col. Hiram I. Bearss commanded the One hundred and second Regiment of the Fifty-first Infantry Brigade, Twenty-sixth Division, in the St. Mihiel offensive. Col. Frederic M. Wise commanded the Fifty-ninth Regiment of the Eighth Infantry Brigade, Fourth Division, from September 5, 1918, to January 4, 1919, during which period he participated in the St. Mihiel and Meuse-Argonne major operations. From January 1, 1919, to February 9, 1919, Col. Wise commanded the Eighth Infantry Brigade of the Fourth Division.

A few Marine officers and enlisted men engaged in Army aviation operations and suffered casualties. About 20 Marine officers were sent to France as observers and as such participated in operations with American, French, and British forces. Marine aviation personnel served in France as the Day Wing of the Northern Bombing Group of the Navy. Marine flyers served with Squadrons 213 (pursuit squadron), 217, and 218 (bombing squadrons), Royal Flying Corps of England; and with pursuit, observation, and bombing squadrons of the French Flying Corps. Quite a few casualties were suffered by the Marine aviation personnel.

The First Marine Aeronautic Company, naval base No. 13, Ponta Delgada, Azores, equipped for water flying only, performed patrol duty from January, 1918, until November 11, 1918.

The Marine Aviation Section, naval air station, Miami, Fla., performed arduous patrol duties in the Florida Straits in connection with the Navy from July, 1918, until the date the armistice went into effect.

Marine detachments served on board all the American battleships attached to the British Grand Fleet and also on the American battleships which based at Castletown Berehaven, Bantry Bay, Ireland. Marines also served on board many of the cruisers which escorted the vessels transporting Army troops to Europe. They were also attached to many other naval vessels such as the *Brooklyn*, *Helena*, and *Wilmington*, in China and Siberian waters, at one time landing at Vladivostok in conjunction with other naval forces; on the *Galveston* on the Murman Coast; and on the *Pittsburgh* in South American waters. Marines were also on the *San Diego* when that vessel was sunk, and the *Minnesota* when that ship was damaged by German mines. Marines were in intimate contact with the Germans in Guam and Philadelphia in conjunction with the Navy in the first hours of the war.

One brigade of Marines was held in readiness in Texas for possible trouble in Mexico which might endanger the Allies' oil supply. Another was scattered throughout the island of Cuba. Large detachments of Marines were stationed in the Azores and Virgin Islands in the nature of advanced base forces, while an advanced base force at Philadelphia was available at all times for naval needs.

Marine forces were also stationed in Guam, Philippine Islands, Peking, Pearl Harbor, and Nicaragua and they assisted materially, under the limited conditions, in the war.

Active operations were conducted in Haiti and Santo Domingo against bandits during the period of the war by Marine forces, the Haitian Gendarmerie and the Guardia Nacional Dominicana, the two latter organizations being composed of natives and administered and officered by the Marine and Navy personnel. Casualties were suffered by Marines in the operations in Santo Domingo, 4 Marines being killed, 13 wounded, and 1 officer wounded, between April 6, 1917, and November 11, 1918.

Chapter IX.

UNITS COMPOSING, AND THE COMMANDING GENERALS OF, THE SECOND DIVISION–VERDUN OPERATIONS.

THE SECOND DIVISION OF REGULARS.

The first unit which ultimately formed a part of the Second Division arriving in France was the Fifth Regiment of Marines which landed in France with the first expedition of American troops in June, 1917. One Marine lieutenant colonel, who afterwards was the first chief of staff of the Second Division, and another Marine lieutenant colonel, who later commanded the Seventeenth Field Artillery of the Second Division, accompanied Gen. Pershing and his staff when they sailed from the United States late in May, 1917.

The Second Division was composed of the following units:

- Third Infantry Brigade:
  - Ninth Infantry.
  - Twenty-third Infantry.
  - Fifth Machine Gun Battalion.
- Fourth Infantry Brigade:
  - Fifth Marines.
  - Sixth Marines.
  - Sixth Machine Gun Battalion of Marines.
- Second Field Artillery Brigade:
  - Twelfth Field Artillery.
  - Fifteenth Field Artillery.
  - Seventeenth Field Artillery.
  - Second Trench Mortar Battery.
- Other troops:
  - Second Engineers.
  - Fourth Machine Gun Battalion.
  - First Field Signal Battalion.
  - Second Headquarters Train and Military Police.
  - Second Ammunition Train.
  - Second Engineer Train.
  - Second Supply Train.
  - Second Sanitary Train.

On October 26, 1917, Brig. Gen. Charles A. Doyen, United States Marine Corps, assumed command of the Second Division as its first commanding general and announced his staff in General Orders, No. 1, with station at Bourmont, Haute-Marne, France. Lieut. Col. Logan Feland, United States Marine Corps, was the first chief of staff. On November 8, 1917, Maj. Gen. Omar Bundy, United States Army, assumed command, published such fact in General Orders, No. 4, November 8, 1917, and was in command of it during the operations in the Verdun and Chateau-Thierry sectors. Maj. Gen. James G. Harbord, United States Army, commanded the division in the Aisne-Marne (Soissons) offensive in July, 1918. Maj. Gen. John A. Lejeune assumed command of the division on July 28, 1918, and retained command until its demobilization in August, 1919. Many Marine officers occupied positions of importance and responsibility on the staff of the commanding general, Second Division. A Marine officer commanded the Seventeenth Field Artillery during the Meuse-Argonne offensive, and other Marine officers commanded battalions of the Ninth Infantry and Fifteenth Field Artillery for a time.

Neither the Marine Brigade nor any other element of the Second Division was the first American unit to enter the front lines since the First Division enjoyed that honor in October, 1917, when it entered the line in the quiet Toul sector.

The Fourth Brigade remained in the Bourmont training area, with headquarters at Damblain, until March 14, 1918, when it commenced movement into subsectors of the Verdun front, the first units of the brigade entering the front line during the night of March 16-17, 1918, with headquarters at Toulon. On April 1, 1918, brigade headquarters was changed to Moscou. The brigade remained on the Verdun front until May 14, 1918, when it proceeded to an area around Vitry-le-Francois for open warfare training, with headquarters at Venault-les-Dames. In the meantime, on May 6, 1918, Brig. Gen. James G. Harbord assumed command of the brigade, relieving Brig. Gen. Doyen who had been ordered to the United States on account of his physical condition. Brig. Gen. Doyen relinquished command of the brigade most unwillingly, and the reasons for his relief are best set forth in the words of the citation of a Navy distinguished service medal posthumously awarded to him, reading as follows:

> By reason of his abilities and personal efforts, he brought this brigade to the very high state of efficiency which enabled it to successfully resist the German army in the Chateau-Thierry sector and Belleau Woods. The strong efforts on his part for nearly a year undermined his health and necessitated his being invalided to the United States before having the opportunity to command the brigade in action, but his work was shown by the excellent service rendered by the brigade, not only at Belleau Woods, but during the entire campaign when they fought many battles.

Gen. Pershing in a letter to Brig. Gen. Doyen stated in part:

> Your service has been satisfactory and your command is considered as one of the best in France. I have nothing but praise for the service which you have rendered in this command.

On May 14, 1918, the brigade left the area around Vitry-le-Francois as it was unsuitable and proceeded to an area around Gisors-Chaumont-en-Vixen, with headquarters at Bou-des-Bois. The brigade was in this area when sudden orders came to move to the Chateau-Thierry sector.

On May 27, 1918, Brig. Gen. John A. Lejeune and Maj. Earl H. Ellis sailed from New York on board the *Henderson* and arrived at Brest, France, on June 8, 1918.

Chapter X.

AISNE DEFENSIVE, HILL 142, BOURESCHES, AND BOIS DE LA BRIGADE DE MARINE, IN THE CHATEAU-THIERRY SECTOR.

In order to appreciate understandingly the importance of the early operations participated in by the Marine Brigade as a unit of the Second Division it is necessary to remember that in 1918, prior to the middle of July, the offensive was in the hands of the Imperial German Staff, and that between March 21, 1918, and July 15, 1918, the Germans directed no less than five major offensives against the Allied lines in efforts to bring the war to a successful conclusion for the Central Powers. American troops assisted in breaking up every one of these drives, but the Second Division, including the Marines, opposed only one, that in the Chateau-Thierry sector. It should also be noted that on March 28, 1918, the American commander in chief placed all of the American forces at the disposal of Marshal Foch, who had been agreed upon as commander in chief of the Allied Armies, to be used as he might decide.

The first offensive (Somme) of the Germans was stopped within a few miles of Amiens, and the second (Lys) overran Armentieres. In this second German offensive, which lasted from April 9 to 27, 1918, and which has been designated by the Americans as a major operation, there were approximately 500 American troops engaged.

Then late in May, 1918, with startling success, which brought a corresponding depression to the morale of the Allies, the Germans launched their third offensive, west of Rheims, crossed the Chemin-des-Dames, captured Soissons, and the last day of May found them marching in the direction of Paris down the Marne Valley. Again the American commander in chief placed every available man at the disposal of Marshal Foch. It was at this critical time, when the Allies were facing a grave crisis, that the Second Division, including the Marine Brigade, together with elements of the Third and Twenty-eighth Divisions, were thrown into the line and, in blocking the German advance in the Chateau-Thierry sector, rendered great assistance in stopping the most dangerous of the German drives.

The first report of the American commander in chief states that "the Third Division, which had just come from its preliminary training area, was hurried to the Marne. Its motorized machine-gun battalion preceded the other units and successfully held the bridgehead at the Marne opposite Chateau-Thierry. The Second Division, in reserve near Montdidier, was sent by motor trucks and other available transport to check the progress of the enemy toward Paris."

The final report of the American commander in chief with reference to this third German offensive stated in part:

> On reaching the Marne that river was used as a defensive flank and the German advance was directed toward Paris. During the first days of June something akin to a panic seized the city and it was estimated that 1,000,000 people left during the spring of 1918. * * *

The Second Division, then in reserve northwest of Paris and preparing to relieve the First Division, was hastily diverted to the vicinity of Meaux on May 31, and, early on the morning of June 1, was deployed across the Chateau-Thierry-Paris road near Montreuil-aux-Lions in a gap in the French line, where it stopped the German advance on Paris.

Without minimizing in any way the splendid actions of the Twenty-sixth Division at Seicheprey and Xivray in April 1918, or the brilliant exploit of the First Division at Cantigny on May 28, 1918, the fact remains that the Second Division, including the Marine Brigade, was the first American division to get a chance to play an important part on the western front, and how well it repelled this dangerous thrust of the Germans along the Paris-Metz highway is too well known to be dwelt upon at length in this brief history.

The fighting of the Second Division in the Chateau-Thierry sector was divided into two parts, one a magnificently stubborn defensive lasting a week and the other a vicious offensive. The defensive fighting of the Second Division between May 31 and June 5, 1918, was part of the major operation called by the Americans the Aisne defensive. Without discussing at this time the tactical or strategical significance of the work of the Second Division in the Aisne defensive, suffice to say that its psychological effect upon the morale of the Allies was tremendous and has been recognized in practically every writing worthy of consideration up to the present date.

The close of the Aisne defensive on June 5, 1918, found the line of the Second Division well established at that point of the Marne salient nearest Paris, but not including Hill 142, Bois de Belleau, Bouresches, or Vaux, and the Germans were in possession of Chateau-Thierry on the right of the Second Division, and continued to hold that town until about July 17, 1918.

On June 6, 1918, the Second Division snatched the initiative from the Germans and started an offensive on its front which did not end until July 1, 1918. The Marine Brigade captured Hill 142 and Bouresches on June 6, 1918, and in the words of Gen. Pershing, "sturdily held its ground against the enemy's best guard divisions," and completely cleared Bois de Belleau of the enemy on June 26, 1918, a major of Marines sending in his famous message: "Woods now U. S. Marine Corps' entirely." The American commander in chief in his first report calls this fighting "the battle of Belleau Wood" and states, "our men proved their superiority, and gained a strong tactical position with far greater loss to the enemy than to ourselves." In his final report he states: "The enemy having been halted, the Second Division commenced a series of vigorous attacks on June 4, which resulted in the capture of Belleau Woods [on June 26] after very severe fighting. The village of Bouresches was taken soon after [on June 6] and on July 1 Vaux was captured. In these operations the Second Division met with most desperate resistance by Germany's best troops." On July 1, 1918, the Third Brigade captured Vaux. The Artillery, Engineers, and the other elements of the Second Division assisted materially in these successes, while the Seventh regiment of the Third Division was in Belleau Wood for a few days about the middle of June.

During these 31 days of constant fighting, the last 26 of which has been defined by general headquarters of the American Expeditionary Forces as a "local engagement," the Second Division suffered

1,811 battle deaths (of which approximately 1,062 were Marines) and suffered additional casualties amounting to 7,252 (of which approximately 3,615 were Marines). It was that fighting and those 9,063 casualties that first made the name Chateau-Thierry famous.

The achievements of the Fourth Brigade of Marines in the Chateau-Thierry sector was twice recognized by the French. The first, which changed the name of the Bois de Belleau, was a beautiful tribute spontaneously made to the successes and to the losses of the Fourth Brigade of Marines, and shows the deep effect that the retaking of Belleau Wood and other near-by positions from the Germans had on the feelings of the French and the morale of the Allies. Official maps were immediately modified to conform with the provisions of the order, the *plan directeur* used in later operations bearing the name "Bois de la Brigade de Marine." The French also used this new name in their orders, as illustrated by an ordre général dated August 9, 1918, signed by the commanding general of the Sixth French Army, reading in part as follows:

Avant la grande offensive du 18 Juillet, les troupes américaines faisant partie de la VIe Armée francaise se sont distinguées en enlevant á l'ennemi le Bois de la Brigade De Marine et le village de Vaux, en arretant son offensive sur la Marine et á Fossoy.

The order changing the name of Bois de Belleau reads as follows:

VI° ARMEE, ETAT-MAJOR,
au Q. G. A., le 30 Juin, 1918.

6930/2.]

ORDRE.

En raison de la brillante conduite de la 4éme Brigade de la 2éme D. U. S. qui a enlèvé de haute lutte Bouresches et le point d'appui important du Bois de Belleau, défendu avec acharnement par un adversaire nombreux, le général commandant la VI° Armee décide que dorénavant, dans tôutes les piéces officielles, le Bois de Belleau portera le nom de "Bois de la Brigade de Marine."

Le Général de Division Degoutte,
Commandant la VI° Armee.
(Signed) DEGOUTTE

A. M. le GÉNÉRAL CDT. la 4ME BRIGADE de MARINE.
s/c. de M. le Général Cdt. la 2me D. U. S.

The second recognition by the French of the Marines' work in the Chateau-Thierry sector were citations of the Fourth Brigade, Fifth and Sixth Regiments, and the Sixth Machine Gun Battalion of Marines, in French army orders, that of the brigade, the others being identical, reading as follows:

Aprés approbation du genéral commandant en chef les forces expéditionnaires américaines en France, le général commandant en chef les armées francaises du nord et du nord-est, cite á l'Ordre de l'Armee:

"4° *Brigade Americaine* sous les ordres du Général de Brigade James G. Harbord, comprenant: Le 5e Regiment de Marine, sous les ordres du Colonel Wendell C. Neville, le 6e régiment de Marine, sous les ordres du Colonel Albertus W. Catlin, le 6e Batailion de mitrailleuses, sous les ordres du Commandant Edward B. Cole:

"A été jetée en pleine bataille, sur un front violemment attaqué par l'ennemi. S'est affirmée aussitôt comme une unité de tout premier order. Dés son entrée en ligne, a brisé, en liaison avec les troupes francaises, une violente attaque ennemie sur un point important de la position et entrepris ensuite á son compte une série d'opérations offensives. Au cours de ces opérations, grace au courage brillant, á la vigueur, á l'allant, á la ténacité de ses hommes qui ne se sont laissés rebuter ni par les fatigues, ni par les pertes; grace á l'activité et á l'énergie de ses officiers; grace enfin á l'action personnelle de son chef, le Général J. Harbord, la 4e brigade a vu ses efforts couronnés de succés. En intime liaison l'un avec l'autre, ses deux régiments et son bataillon de mitrailleuses ont réalisé, après douze jours de lutte incessante (du 2 au

13 Juin 1918) dans un terrain trés difficile, une progression variant entre 1,500 á 2,000 métres, sur un front de 4 kilométres, capturant un nombreux matériel, faisant plus de 500 prisonniers, infligeant á l'ennemi des pertes considérables et lui enlevant deux points d'appui de premiére importance—le village de Bouresches et le bois organisé de Belleau."

Au Grand Quartier Général, le 22 octobre, 1918.

Le Général Commandant en Chef.
Signé: PETAIN

(Ordre No. 10.805 "D.")

In addition to the above-described instances, French civilian sentiment expressed itself in the following letter from the mayor of Meaux and Resolution from the assembled mayors of the Meaux District (Arrondissement). This letter and the resolutions were published on July 10, 1918, in General Orders No. 43, of the Second Division "as indicating the appreciation of the efforts of the Second Division by the French inhabitants for our share in stemming the recent German advance in this sector."

MEAUX, *June 26, 1918.*

GENERAL: On behalf of all the Mayors of the Meaux District (Arrondissement), assembled yesterday in congress at the city hall, I have the honor to send you herewith a copy of the resolution they have taken in order to pay homage to the gallantry displayed by the troops under your command and to the effectiveness of the help they rendered us.

The civilian population of this part of the country will never forget that the beginning of this month of June, when their homes were threatened by the invader, the Second American Division victoriously stepped forth and succeeded in saving them from impending danger.

I am personally happy to be able to convey to you this modest token of their thankfulness and I am, General,

Yours, respectfully,

(Signed) G. LUGOL,
Mayor of Meaux, Depute de Seine et Marne.

Voted in a Congress of the Mayors of Meaux District on the 25th of June, 1918.

The mayors of the Meaux district, who were eye-witnesses to the generous and efficacious deeds of the American Army in stopping the enemy advance, send to this Army the heart-felt expression of their admiration and gratefulness.

(Signed) G. LUGOL,
President of the Committee.

MEAUX, *June 25, 1918.*

During the first attack on Belleau Wood on June 6, 1918, Col. Albertus W. Catlin was severely wounded and was relieved in command of the Sixth Regiment by Lieut. Col. Harry Lee, who continued in command until the regiment was demobilized in August, 1919.

When Maj. Edward B. Cole was mortally wounded on June 10, 1918, he was relieved in command of the Sixth Machine Gun Battalion of Marines by Capt. Harlan E. Major. On June 11, 1918, Captain Major was relieved by Capt. George H. Osterhout, who retained command until relieved by Maj. Littleton W. T. Waller, jr., on June 21, 1918.

During the fighting in the Chateau-Thierry sector the headquarters of the Fourth Brigade was successively at Montreuil-aux-Lions, (in an automobile for one-half hour on the way to the front lines), Issonge farmhouse, and La Loge farmhouse. After being relieved by elements of the Twenty-sixth Division during the night of July 5–6, 1918, the brigade moved to an area in rear of the lines and occupied what was known as the Line of Defense or Army Line, with headquarters at Nanteuil-sur-Marne. The brigade remained there until July 16, 1918.

During the time the above-described fighting was going on the Germans were frustrated in their fourth 1918 drive (Noyon-Montdidier defensive) between June 9 and 15, 1918, and of course being busy in the vicinity of Bois de Belleau, the Marines had no opportunity of engaging in it.

Having been blocked in the Marne salient, the Germans attacked for the fifth time in 1918 on July 15, and as events turned out it was the last, for from the time of its failure they were on the defensive. The Allied troops including many Americans held this attack, called by the Americans the Champagne-Marne defensive, which was on a large scale, and the grand initiative passed from the Germans to the Allies on July 18, 1918, when Marshal Foch launched his initial major offensive, termed by the Americans the Aisne-Marne. In this magnificent and gigantic operation the Marine Brigade and other elements of the Second Division played leading parts in the vicinity of Soissons.

General headquarters, American Expeditionary Forces, on May 28, 1919, credited the Second Division units with participation in the major operation of Champagne-Marne defensive, but on June 2, 1919, rescinded this credit.

Chapter XI.

THE AISNE-MARNE OFFENSIVE (SOISSONS).

On July 11, 1918, Brig. Gen. James G. Harbord, commanding general of the Marine Brigade, received notification of his appointment as a major general, and two days later left on a five days' leave of absence. As Col. Neville had been evacuated to a base hospital after leaving the Chateau-Thierry sector, Lieut. Col. Harry Lee assumed temporary command of the brigade. Maj. Gen. Harbord and Col. Neville both returned in time to enter the Aisne-Marne offensive, the former in command of the Second Division and the latter in command of the Fourth Brigade.

Of the six Allied offensives taking place in 1918 on the Western Front, designated by the Americans as major operations, the Fourth Brigade of Marines, with the other units of the Second Division, participated in three, the first being the vast offensive known as the Aisne-Marne, in which the Marine Brigade entered the line near Soissons.

On July 17, 1918, the First Moroccan Division and the First and Second Divisions of American Regulars were hurriedly and secretly concentrated, by terribly fatiguing, forced night marches over roads jammed with troops, artillery, and tanks, through rain and mud, in the Bois de Retz, near Soissons. Headquarters of the Fourth Brigade was established at Vivieres.

The getting to the "jump-off" on time for this operation will always share in Marine Corps history with the glorious victory that followed.

Early on the morning of July 18, 1918, Marshal Foch threw these three picked divisions at the unsuspecting Germans with overwhelming success, and again on the following day. The American commander in chief in his first report stated:

> The place of honor in the thrust toward Soissons on July 18 was given to our First and Second Divisions, in company with chosen French divisions. Without the usual brief warning of a preliminary bombardment, the massed French and American artillery, firing by the map, laid down its rolling barrage at dawn while the Infantry began its charge. The tactical handling of our troops under these trying conditions was excellent throughout the action. * * * The Second Division took Beaurepaire Farm and Vierzy in a very rapid advance, and reached a position in front of Tigny at the end of its second day.

In his final report he stated:

> Gen. Petain's initial plan for the counterattack involved the entire western face of the Marne salient. The First and Second American Divisions, with the First French Moroccan Division between them, were employed as the spearhead of the main attack, driving directly eastward, through the most sensitive portion of the German lines to the heights south of Soissons. The advance began on July 18, without the usual brief warning of a preliminary bombardment, and these three divisions at a single bound broke through the enemy's infantry defenses and overran his artillery, cutting or interrupting the German communications leading into the salient. A general withdrawal from the Marne was immediately begun by the enemy, who still fought stubbornly to prevent disaster. * * *

The Second Division advanced 8 kilometers in the first 26 hours, and by the end of the second day was facing Tigny, having captured 3,000 prisoners and 66 field guns. It was relieved the night of the 19th by a French division. The result of this counter-offensive was of decisive importance. Due to the magnificent dash and power displayed on the field of Soissons by our First and Second Divisions the tide of war was definitely turned in favor of the Allies.

Maj. Gen. James G. Harbord, commanding the Second Division in this operation, describes the two days' fighting of his division in these words:

It is with keen pride that the division commander transmits to the command the congratulations and affectionate personal greetings of Gen. Pershing who visited the division headquarters last night. His praise of the gallant work of the division on the 18th and 19th is echoed by the French high command, the Third Corps commander, American Expeditionary Forces, and in a telegram from the former division commander. In spite of two sleepless nights, long marches through rain and mud, and the discomforts of hunger and thirst, the division attacked side by side with the gallant First Moroccan Division and maintained itself with credit. You advanced over 6 miles, captured over 3,000 prisoners, 11 batteries of artillery, over 100 machine guns, minnenwerfers, and supplies. The Second Division has sustained the best traditions of the Regular Army and the Marine Corps. The story of your achievements will be told in millions of homes in all Allied lands to-night.

This was one of the greatest strategical successes of Marshal Foch, and that the part played by the Marines was appreciated by the French is illustrated by the Fifth and Sixth Regiments and the Sixth Machine Gun Battalion being cited in French Army orders. The citations of the Sixth Regiment (that of the Fifth Regiment being similar) and that of the Sixth Machine Gun Battalion are quoted below:

Aprés approbation du général commandant en chef les forces expéditionnaires Américaines en France, le général commandant en chef les armées Françaises du nord et du nord-est, cite á l'Ordre de l'Armée:

"Le 6e Regiment de Marine Americaine, sous les ordres du Lieutenant-Colonel Lee,

"Engagés á l'improviste dans l'offensive du 18 juillet 1918, en pleine nuit, dans un terrain inconnu et trés difficile, ont déployé pendant deux jours, sans se laisser arrêter par les fatigues et les difficultés du ravitaillement en vivres et en eau, une ardeur et une ténacité remarquables, refoulant l'ennemi sur 11 kilométres de profondeur, capturant 2,700 prisonniers, 12 canons et plusieurs centaines de mitrailleuses."

Au Grand Quartier Général, le 25 Octobre 1918.

(Ordre No. 10.886 "D.")

Le Général Commandant en Chef.
Signé: PETAIN

Aprés approbation du général commandant en chef les forces expéditionnaires Américaines en France, le maréchal de France, commandant en chef les Armées Françaises de l'est cite á l'Ordre de l'Armée:

"Le 6e Bataillon de Mitrailleuses U. S. Marine, sous les ordres du Commandant L. W. T. Waller.

"Quoique trés fatigué par un long trajet en camion et une marche de nuit sur des routes difficiles, ce bataillon s'est précipité á l'attaque le 18 juillet 1918, prés de Vierzy et a puissamment contribué á consolider et á maintenir la position atteinte ce jour-lá.

"Dans la matinée du 19 juillet, il s'est vaillamment porté en avant, en terrain découvert, sous un violent feu d'artillerie et de mitrailleuses, soutenant résolument l'attaque lancée contre les positions renforcées de l'ennemi.

"Ayant á faire face á une forte résistance ennemie et á des contre-attaques continuelles, a fait preuve du plus beau courage en consolidant rapidement et en tenant résolument l'importante position conquises par l'infanterie ce jour-lá."

Au Grand Quartier Général, le 4 Mars 1919.

(Ordre No. 13.978 "D.")

Le Maréchal,
Commandant en Chef les Armées Françaises de l'Est.
PETAIN

Following the advance of the first day, brigade headquarters was moved forward to a cave in Vierzy.

Col. Logan Feland was in command of the Fifth Regiment during the Aisne-Marne offensive, near Soissons, and continued in command of it with the exception of two days in July, 1918 (when Brig. Gen. Lejeune commanded the Fourth Brigade and Col. Neville the Fifth Regiment), until March 21, 1919, when he was relieved by Col. Harold C. Snyder, who retained command until the date of demobilization.

The Fourth Brigade was relieved about midnight July 19, 1918, and after remaining in a reserve position until July 22, 1918, marched to an area farther in the rear, but still in a reserve position, brigade headquarters being established at Taillefontaine. After final relief from this active sector the brigade was billeted July 24–25, 1918, in an area around Nanteuil-le-Haudouin, brigade headquarters being established at Nanteuil. The brigade remained in this area until July 31, 1918.

On July 25, 1918, Brig. Gen. John A. Lejeune arrived, and assumed command of the Fourth Brigade on July 26, 1918, General Orders, No. 16, reading as follows:

> I have this day assumed command of the Fourth Brigade, U. S. Marines.
>
> To command this brigade is the highest honor that could come to any man. Its renown is imperishable and the skill, endurance, and valor of the officers and men have immortalized its name and that of the Marine Corps.

Brig. General Lejeune retained command until July 29, 1918, when he became commanding general of the Second Division, relieving Maj. Gen. Harbord, who left to assume command of the Services of Supply. Col. Neville, on this latter date, resumed command of the Fourth Brigade.

Chapter XII.

MARBACHE SECTOR, NEAR PONT-A-MOUSSON—ST. MIHIEL OFFENSIVE.

During the last two days of July, 1918, the units of the brigade entrained for a 24-hour railroad journey which took them to an area around Nancy, with headquarters at Villers-les-Nancy, where they remained resting and refitting until August 9, 1918.

On August 7, 1918, information was received of the promotion of Brig. Gen. Lejeune to the grade of major general, and of Col. Neville to the grade of brigadier general, both to date from July 1, 1918.

Col. Albertus W. Catlin arrived in the United States on board the *America* on August 3, 1918. Col. Catlin, having been wounded on June 6, 1918, during the first attack on Bois de Belleau, was admitted to Hospital No. 2, Paris, France, on the next day, was discharged on July 22, 1918, granted two months' sick leave, and sailed for New York from Brest, France, on July 25, 1918.

On August 5, 1918, movement of units of the brigade was started for the occupation of the Marbache subsector, near Pont-a-Mousson, on the Moselle River. By August 8, 1918, the movement was completed, with Headquarters established at Scarponne just across the Moselle River from Dieulouard. The sector was quiet and occupation uneventful except for an enemy raid which was successfully repulsed and prisoners captured.

On August 8, 1918, Lieut. Col. Earl H. Ellis was appointed adjutant of the Fourth Brigade, relieving Lieut. Col. Harry R. Lay, who had been detailed as inspector general of the Second Division.

The relief from the Marbache sector was completed on August 18, 1918, and the brigade moved to an area about 20 kilometers southeast of Toul, headquarters being established at Favieres. Intensive training for the impending St. Mihiel offensive was indulged in here.

The brigade started to move from this area on the night of September 2, 1918, and after a series of night marches, during which time headquarters were established at Pont St. Vincent, Velaine-en-Haye, and Bouvron, the brigade arrived just outside of Manonville, headquarters being established in Manonville. From September 12 to 16, 1918 the brigade was engaged in the St. Mihiel offensive in the vicinity of Remenauville, Thiaucourt, Xammes, and Jaulny as a unit of the Second Division of the First Corps of the First Army. Headquarters during these operations were successively at 1 kilometer north of Lironville, Thiaucourt, and finally at Manonville, on September 16, 1918.

On September 20, 1918, the brigade moved to an area south of Toul, with headquarters at Chaudenay. The brigade remained in this area until September 25, 1918, when it moved by rail to an area south of Chalons-sur-Marne, with headquarters at Sarry.

Chapter XIII.

THE CHAMPAGNE—BATTLE OF BLANC MONT RIDGE—CAPTURE OF ST. ETIENNE—MARCH TO LEFFINCOURT.

Marshal Foch, having asked for an American division to assist in breaking through the powerful German defenses in the Champagne, the Second Division, including the Marine Brigade, was temporarily placed at the disposal of the Fourth French Army under Gen. Gouraud from September 27, 1918, to October 10, 1918. At first it was directly subject to the orders of Marshal Petain, but before the actual fighting began it was placed directly under the orders of Gen. Gouraud.

On September 28, 1918, the Fourth Brigade moved by bus and marching to the Souain-Suippes area, with brigade headquarters at Suippes.

On October 1, 1918, in an order of the Second Division, the commanding general of the Second Division encouraged his division with the following words:

> 1. The greatest battles in the world's history are now being fought. The Allies are attacking successfully on all fronts. The valiant Belgian Army has surprised and defeated the enemy in Flanders; the English, who have been attacking the enemy without ceasing since August 8, have advanced beyond the Hindenburg Line, between Cambria and St. Quentin, capturing thousands of prisoners and hundreds of cannon; the heroic Allied Army of the Orient has decisively defeated the Bulgars; the British have captured over 50,000 prisoners in Palestine and have inflicted a mortal blow on the Turk; and our own First Army and the Fourth French Army have already gained much success in the preliminary stages of their attack between the Meuse and Suippes Rivers.
>
> 2. Owing to its world-wide reputation for skill and valor, the Second Division was selected by the commander in chief of the Allied Armies as his special reserve, and has been held in readiness to strike a swift and powerful blow at the vital point of the enemy's line. The hour to move forward has now come, and I am confident that our division will pierce the enemy's line, and once more gloriously defeat the Hun.

The Battle of Blanc Mont Ridge was one of the most powerful and effective blows struck under the direction of Marshal Foch against the retreating Germans, and its brilliantly successful conclusion was due in a great degree to the military genius of Maj. Gen. John A. Lejeune of the Marines.

On September 27, 1918, Maj. Gen. John A. Lejeune called on Gen. Gouraud at the headquarters of the Fourth French Army, who explained the situation at the front to him. Facing a large relief map of the battlefield, Gen. Gouraud placed his hand on the Blanc Mont Ridge and said: "General, this position is the key of all the German defenses of this sector including the whole Rheims Massif. If this ridge can be taken the Germans will be obliged to retreat along the whole front 30 kilometers to the river Aisne. Do you think your division could effect its capture?" Maj. Gen. Lejeune responded that he felt certain the Second Division could take the stronghold pointed out, whereupon he was informed that he would be ordered to make the attack within a few days and was directed to prepare a plan for the assault.

At this time the Second Division was directly subject to the orders of Marshal Petain, but later in the day Gen. Gouraud informed Maj. Gen. Lejeune that after an explanation of the circumstances Marshal Petain had assigned the division to the Fourth French Army.

The general plan provided for an attack by the whole Fourth French Army between the Argonne and the Suippes River.

On October 1, 1918, the brigade with the rest of the Second Division marched to the front line near Somme-Py on the night of October 1–2, 1918, and relieved elements of a French division. The brigade headquarters was located in the trenches about 2½ kilometers south of Somme-Py. The relief was effected before daylight without incident.

The Battle of Blanc Mont Ridge was fought and won by the Second Division, as a unit of the Fourth French Army, between October 3 and 9, 1918, over the desolated white chalky ground of the Champagne, which was scarred and shell pocked by years of artillery fire, marked with huge mine craters, gridironed with an intricate maze of deep trenches and concrete fortifications, and covered with tangled masses of wire.

The overwhelming success and the far-reaching effect of the Second Division's part in these operations, the cleaning up of the Essen Hook, the capture of Blanc Mont Ridge, and the capture of St. Etienne, are well described in general terms in the following excerpts from official publications.

That the plan was as brilliantly executed as it was daringly conceived is shown by this extract from an order of the Second Division, dated November 11, 1918, reading in part as follows:

> In the Champagne district, October 2 to 10, it fought beside the Fourth French Army. On October 3 it seized Blanc Mont Ridge, the keystone of the arch of the main German position, advanced beyond the ridge and, although both flanks were unsupported, it held all its gains with the utmost tenacity, inflicting tremendous losses on the enemy. This victory freed Rheims and forced the entire German Army between that city and the Argonne Forest to retreat to the Aisne, a distance of 30 kilometers.

The amazing success of the attack and the vital effect of the capture of Blanc Mont Ridge and St. Etienne is described in the words of Gen. Gouraud himself in a letter to Marshal Foch, reading in part as follows:

> Because of the brilliant part played by this "Grand Unit" in the offensive of the Fourth Army during the autumn of 1918, I propose the Second American Division for a citation in "The Order of the Army" upon the following specific grounds:
>
> The Second Infantry Division, United States, brilliantly commanded by Gen. Lejeune * * * played a glorious part in the operations of the Fourth Army in the Champagne in October, 1918. On the 3d of October this division drove forward and seized in a single assault the strongly entrenched German positions between Blanc Mont and Medeah Ferme, and again pressing forward to the outskirts of Saint Etienne-a-Arnes it made in the course of the day an advance of about 6 kilometers.
>
> It captured several thousand prisoners, many cannon and machine guns, and a large quantity of other military matériel. This attack, combined with that of the French divisions on its left and right, resulted in the evacuation by the enemy of his positions on both sides of the river Suippe and his withdrawal from the Massif de Notre-Dame-des-Champs.

The further opinion of the French as to the results and effect of the Second Division's operations in Champagne is set forth in the following-quoted extract from Information Bulletin No. 12 of the Fourth French Army dated October 7, 1918:

Up to October 4, at which date the present bulletin is written, the Fourth Army has pushed its advance up to objectives of the very highest importance. A splendid American division, full of dash and ardor, the Second Division, United States, placed at the disposition of the Twenty-first Corps on October 3, made itself master of Massif du Blanc Mont, which dominates the valley of the Arnes and gives us excellent outlook on the valley of the Suippe in rear of the region of Monts. This conquest rapidly brought about the downfall of Notre-Dame-des-Champs and the Grand Bois de Saint Souplet.

The American commander in chief in his first report describes the Battle of Blanc Mont in the following words:

The Second Division conquered the complicated defense works on their front against a persistent defense worthy of the grimmest period of trench warfare and attacked the strongly held wooded hill of Blanc Mont, which they captured in a second assault, sweeping over it with consummate dash and skill. This division then repulsed strong counterattacks before the village and cemetery of St. Etienne and took the town, forcing the Germans to fall back from before Rheims and yield positions they had held since September, 1914.

In his final report the American commander in chief remarked as follows:

The Second Division completed its advance on this front by the assault of the wooded heights of Mont Blanc, the key point of the German position, which was captured with consummate dash and skill. The division here repulsed violent counterattacks and then carried our lines into the village of St. Etienne, thus forcing the Germans to fall back before Rheims and yield positions which they had held since September, 1914.

The citation of the Fifth Regiment of Marines (the citation of the Sixth Regiment being identical) reads as follows:

Aprés approbation du général commandant en chef les forces expéditionnaires Americaines en France, le maréchal de France, commandant en chef les armées françaises de l'est, cite à l'Ordre de l'Armée:

"Le 5ème Regiment de Marine Americain, sous les ordres du Colonel Logan Feland:

"A pris une part glorieuse aux opérations engagées par la 4ème Armée en Champagne, en Octobre 1918. Le 3 Octobre 1918, a participé à l'attaque des positions allemandes fortement retranchées entre le Blanc-Mont et la Ferme Medeah, et, poussant de l'avant jusqu'aux abords de Saint-Etienne à Arnes, a réalisé une avance de 6 kilomètres. A fait plusieurs milliers de prisonniers, capturé des canons, des mitrailleuses et un important matériel de guerre. Cette attaque, combinée avec celle des Divisions Françaises, a eu pour conséquence l'évacuation des deux rives de la Suippe et du Massif de Notre-Dame-des-Champs."

Au Grand Quartier Général, le 21 Mars 1919.

Le Maréchal, Commandant en Chef les Armées Françaises de l'Est.

Signe: PÉTAIN.

(Ordre No. 14.712 "D.")

On October 10, 1918, having been relieved from the line in the Blanc Mont sector, the brigade took station in the Suippes-Somme Suippes-Nantivet area and the adjacent camps with headquarters at Suippes, being assigned as Fourth French Army reserve. The brigade remained in this area resting and refitting until October 14, 1918, when, in accordance with orders, it marched to the Vadenay-Bouy-la-Veuve-Dampierre area, north of Chalons-sur-Marne, with headquarters at Bouy. While here orders were received placing the Fourth Brigade provisionally at the disposal of the Ninth French Army Corps to hold a sector in the region Attigny-Voncq-Aisne River.

Accordingly on October 20, 1918, the brigade was temporarily detached from the Second Division and marched to the area Suippes-Nantivet-Somme-Suippes, with headquarters at Suippes. On October 21, 1918, in obedience to orders, the Marines hiked to the vicinity of Leffincourt, where brigade headquarters was established. While about to take over the assigned sector the Fourth Brigade received orders to rejoin the Second Division, which was preparing to enter the Meuse-Argonne offensive. After a hard march these orders were obeyed and brigade headquarters established at Mont Pelier on October 23, 1918.

On October 24, 1918, Maj. Matthew W. Kingman relieved Maj. Littleton W. T. Waller, jr., in command of the Sixth Machine Gun Battalion of Marines, Maj. Waller joining the Second Division staff as division machine gun officer.

Chapter XIV.

THE MEUSE-ARGONNE OFFENSIVE—CROSSING THE MEUSE RIVER.

On October 25, 1918, the brigade moved to the Les Islettes area with brigade headquarters at Camp Cabaud. On the evening of October 26, 1918, it arrived in the area south of Exermont and bivouacked in the woods there that night with brigade headquarters at Exermont. The brigade remained in bivouac in this area until the night of October 30–31, 1918, when it moved forward into line to participate in the immense Meuse-Argonne offensive which had started on September 26, 1918, the Second Division being assigned as a unit of the Fifth Corps.

Relieving elements of the Forty-Second Division, just south of Landres-et-St.-Georges, the Marine Brigade early on the morning of November 1, 1918, jumped-off, following a terrific barrage, for its final operation of the war, the conclusion of which at 11 o'clock on the morning of November 11, 1918, found the Marines firmly established on the heights of the far bank of the Meuse River, after an advance of 30 kilometers.

The splendid work of the Second Division, including the Marines, is described in official reports, and excerpts from some are given below.

In recommending that the Second Division be cited in General Headquarters Orders for its excellent work in the attack of November 1–11, 1918, the commanding general, First Army, wrote on January 16, 1919, in part, as follows:

> 4. In the First Army attack of November 1, 1918, the Second Division was selected and so placed in the battle line that its known ability might be used to overcome the critical part of the enemy's defense. The salient feature of the plan of attack was to drive a wedge through Landres-et-St. Georges to the vicinity of Fosse. It was realized that if the foregoing could be accomplished the backbone of the hostile resistance west of the Meuse would be broken and the enemy would have to retreat to the east of the Meuse. Success in this plan would immediately loosen the flanks of the First Army. The Second Division was selected to carry out this main blow.
>
> 5. The Second Division accomplished the results desired in every particular on the first day of the attack, not only clearing the hostile defenses of Landres-et-St. Georges and the Bois de Hazios but continuing its advance to the vicinity of Fosse, i. e., about 9 kilometers. This decisive blow broke the enemy's defense and opened the way for the rapid advance of the Army.

With reference to the first day's attack, the commanding general, Fifth Army Corps, wrote officially on November 2, 1918, in part as follows:

> The division's brilliant advance of more than 9 kilometers, destroying the last stronghold on the Hindenburg Line, capturing the Freya Stellung, and going more than 9 kilometers against not only the permanent but the relieving forces in their front, may justly be regarded as one of the most remarkable achievements made by any troops in this war. For the first time, perhaps, in our experience the losses inflicted by your division upon the enemy in the offensive greatly exceeded the casualties of the division. The reports indicate moreover that in a single day the division has captured more artillery and machine guns than usually falls to the lot of a command during

several days of hard fighting. These results must be attributed to the great dash and speed of the troops, and to the irresistible force with which they struck and overcame the enemy.

The following citation in Fifth Army Corps General Orders No. 26, dated November 20, 1918, gives a further description of these operations:

> The Second Division, in line at the launching of the attack, broke through the strong enemy resistance, and, leading the advance, drove forward in a fast and determined pursuit of the enemy, who, despite new divisions hastily thrown in, was driven back everywhere on its front. This division drove the enemy across the Meuse, and under heavy fire and against stubborn resistance, built bridges and established itself on the heights. The cessation of hostilities found this division holding strong positions across the Meuse and ready for a continuation of the advance.

An order of the Second Division, dated November 5, 1918, reading in part as follows, tells what occurred subsequent to the first day's attack:

> During the night of November 2–3 the Second Division moved forward overcoming the resistance of the enemy's advanced elements, and at 6 a. m., it attacked and seized the enemy's line of defense on the ridge southeast of Vaux-en-Dieulet.
>
> Late in the afternoon, the enemy, having reorganized his line on the border of Belval Forest, was again attacked and defeated. After nightfall and in a heavy rain, the advanced elements of the division pressed forward through the forest, and occupied a position on the heights south of Beaumont, 8 kilometers in advance of the divisions on our right and left.
>
> During the night of November 4–5, the division again pressed forward, occupied Beaumont and Letanne and threw the enemy on its front across the Meuse.

An order of the Second Division, dated November 12, 1918, describing the historic crossing of the Meuse River on the night before the armistice became operative, reads as follows:

> 1. On the night of November 10, heroic deeds were done by heroic men. In the face of a heavy artillery and withering machine gun fire, the Second Engineers threw two foot bridges across the Meuse and the first and second battalions of the Fifth Marines crossed resolutely and unflinchingly to the east bank and carried out their mission.
>
> 2. In the last battle of the war, as in all others in which this division has participated, it enforced its will on the enemy.

The commanding general of the Fifth Army Corps has this to say about the crossing of the Meuse by the Marines, who were assisted by the Artillery, Engineers, and other troops of the Second Division:

> Especially I desire to commend the division for the crowning feat of its advance in crossing the Meuse River in face of heavy concentrated enemy machine gun fire, and in driving the enemy's troops before it, and in firmly establishing itself upon the heights covering the desired bridgehead. This feat will stand among the most memorable of the campaign.

With reference to the crossing of the Meuse River the American commander in chief reported as follows:

> On the night of November 10, the Fifth Corps forced a crossing of the Meuse against heavy enemy resistance between Mouzon and Pouilly, and advanced to the Inor-Mouzon road with two battalions holding the high ground northwest of Inor.

The general success achieved by the Second Division in the Argonne-Meuse offensive is well described by the words of the order citing Maj. Gen. John A. Lejeune, of the Marines, for an Army distinguished service medal, reading in part as follows:

> In the Meuse-Argonne offensive his division was directed with such sound military judgment and ability that it broke and held, by the vigor and rapidity of execution of its attack, enemy lines which had hitherto been considered impregnable.

During this fighting the headquarters of the Fourth Brigade was successively established at Exermont, one-half kilometer north of Exermont, Sommerance, Bayonville-et-Chennery, Fosse, Belval-Bois-des-Dames, and Beaumont.

The following self-explanatory memorandum was sent out by the Commanding General of the Fourth Brigade:

HEADQUARTERS 4TH BRIGADE,
MARINES, AMERICAN E. F.,
11th November '18—8.40 a. m.

PEACE MEMORANDUM NO. 1.

The following telephone message received from *Surprise* 1 at 8.35 a. m. this morning forwarded for compliance.

8.40 a. m. message from 5th Corps; Armistice signed and takes effect at 11 this morning. Accurate map showing locations of front line elements, including patrols and detachments, will be sent to these Headquarters without delay.

On "the eleventh hour, the eleventh day of the eleventh month, of the year 1918," Brig. Gen. Wendell C. Neville, commanding general of the Fourth Brigade of Marines, published the following tribute to the officers and men of the Fourth Brigade:

Upon this, the most momentous hour in the history of the World War, the undersigned wishes to express to his command his sincere appreciation of their unfailing devotion to duty and their heroic and courageous action during the recent operations.

The time, when the results of our efforts during the past year are shown, is here. The hour has arrived when the convulsion which has shaken the foundations of the civilized world has ceased. The enemy is defeated and the principles of freedom and democracy have triumphed over barbarism and autocracy. We may all feel justly proud of the extent of our participation which has forced the enemy to a cessation of hostilities. It is fitting, at this time, to think of those of our comrades who have fallen on the field of honor and rejoice in the fact that they did not give their lives in vain.

Your display of fortitude, determination, courage, and your ability to fight has upon more than one occasion been a determining factor in making history, and your work has had a direct bearing upon the remarkable chain of events which have this day culminated in such a satisfactory manner. Along the fronts of Verdun, the Marne, the Aisne, Lorraine, Champagne, and the Argonne, the units of the Fourth Brigade Marines have fought valiantly, bravely, and decisively. They have nobly sustained the sacred traditions and have added glorious pages to the already illustrious history of the United States Marine Corps. It is a record of which you may all be proud.

Chapter XV.

MARCH TO THE RHINE—ARMY OF OCCUPATION—SUMMARY OF OPERATIONS OF THE FOURTH BRIGADE.

MARCH TO THE RHINE.

On November 17, 1918, the Second Division commenced its march to the Rhine, passing through Belgium and Luxembourg. The German frontier was reached November 25, 1918, crossed on December 1, 1918, the Rhine reached December 10, 1918, and crossed on December 13, 1918. During this march and up to the time the Fourth Brigade settled down to its occupation duty in Germany, brigade headquarters were successively established at Margut, Bellefontaine, Arlon, Usseldange, Berg, Eppeldorf, Neuerburgh, Waxweiler, Prum, Budesheim, Wiesbaum, Antweiler, Neuenahr, Burgbrohl, Rheinbrohl, and Hönningen.

WITH THE ARMY OF OCCUPATION.

The duties of the Fourth Brigade with the Army of Occupation in Germany were uneventful, the outstanding features being the establishment of a Rhine River patrol, manned and commanded by Marines; an extended visit, inspection, and review by the Secretary of the Navy; and the operation of the Second Division, including the Marines, made about the middle of June, 1919, in which an advanced position was taken as a part of the concentration of the Third Army immediately preceding the signing of the treaty of peace by the Germans.

Headquarters of the Fourth Brigade during the greater part of the occupation of Germany was at Nieder Bieber, while during the last operation when the advanced position was taken, just prior to Germany signing the peace treaty, it was at Herschbach. On the date the treaty was signed the Fifth Regiment, with headquarters at Hatenfels, occupied the most advanced position ever occupied by Marines in Germany.

Just before departing from Germany headquarters was at Nieder Bieber, and with the exception of Brest, France, this was the last headquarters the brigade had in Europe.

Maj. Charles D. Barrett relieved Lieut. Col. Earl H. Ellis as brigade adjutant in April, 1919, and held that position until the brigade was demobilized. Lieut. Col. Ellis was assigned to duty as second in command of the Fifth Regiment. On March 12, 1919, Col. Logan Feland was temporarily appointed brigadier general to rank from March 9, 1919, and accepted appointment and executed oath on March 17, 1919. On March 21, 1919, Col. Harold C. Snyder assumed command of the Fifth Regiment, relieving Brig. Gen. Logan Feland, who, after acting as aide for the Secretary of the Navy, arrived in the United States on the *Von Steuben* on May 13, 1919.

Just before the Second Division left Germany the commanding general of the Army of Occupation expressed his appreciation of the services of that division in a letter dated July 2, 1919, which is published in General Orders, Second Division, No. 68, July 5, 1919:

As your magnificent division is about to leave his command, it is with a sense of gratitude for its splendid achievments while in the American Expeditionary Forces that the army commander expresses to you, and to your gallant officers and men his appreciation of your services.

After occupying a defensive sector between Verdun and St. Mihiel, you were placed in the line of battle and met, with stubborn resistance, the onslaughts of the enemy's hordes near Chateau-Thierry. Your action at Belleau Woods and your attack upon and capture of Vaux must ever remain brilliant exploits in our military history.

At Soissons, side by side with a veteran French division, you proved to our Allies the fighting value of the Army of the United States, and at St. Mihiel, in the first great American offensive, your prowess in attack was irresistible.

When in October, 1918, the Allied High Command desired to reinforce the French Army by American troops of great offensive worth, by real "shock troops," you were loaned to Gen. Gouraud's Fourth French Army and delivered your famous assault on Blanc Mont Ridge, releasing from German menace the historic city of Rheims.

In the closing phase of the Meuse-Argonne operations, certainly no troops contributed more to the enemy's destruction than your division. After taking Landres-et-St. Georges, Bayonnville-et-Chennery, and the Bois-de-la-Folie, you pierced the Bois-de-Belval, and by skillful night fighting and marching you cleared the enemy from the left bank of the Meuse and forced a crossing of the river.

Your brilliant exploits in battle are paralleled by the splendid example of soldierly bearing and discipline set by your officers and men while a part of the Army of Occupation. That spirit and dash which carried your men through the enemy's defenses still predominated when the Army was recently concentrated, preparatory to a further advance into unoccupied Germany.

Officers and soldiers of the Second Division, your achievments and sacrifices have earned for you and for your fallen comrades the praise and gratitude of our Nation.

SUMMARY OF THE OPERATIONS OF THE FOURTH BRIGADE.

A summary of the operations of the Fourth Brigade of Marines is set forth below:

The Fourth Brigade of Marines as a unit of the Second Division participated in actual battle in France in the following sectors between the inclusive dates set down (as published in General Orders No. 87, Second Division, April 25, 1919):

Toulon sector, Verdun: From March 15 to May 13, 1918.

Aisne defensive, in the Chateau-Thierry sector: From May 31 to June 5, 1918.

Chateau-Thierry sector (capture of Hill 142, Bouresches, Belleau Wood): From June 6 to July 9, 1918.

Aisne-Marne (Soissons) offensive: From July 18 to July 19, 1918.

Marbache sector, near Pont-a-Mousson on the Moselle River: From August 9 to August 16, 1918.

St. Mihiel offensive, in the vicinity of Thiaucourt, Xammes, and Jaulny: From September 12 to September 16, 1918.

Meuse-Argonne (Champagne) including the capture of Blanc Mont Ridge and St. Etienne: From October 1 to October 10, 1918.

Meuse-Argonne (including crossing of the Meuse River): From November 1 to November 11, 1918.

SILVER BANDS FOR COLORS.

Under the rulings of General Headquarters, American Expeditionary Forces, the Marine Corps units serving with the Second Division are entitled to silver bands on the staffs of their colors for battle participation in the above mentioned engagements.

MAJOR OPERATIONS.

General Headquarters, American Expeditionary Forces, ruled that the Second Division, including the Fourth Brigade of Marines, participated in only four major operations, the Aisne defensive (May 31 to June 5, 1918); the Aisne-Marne offensive (July 18 and 19, 1918); the St. Mihiel offensive (Sept. 12 to 16, 1918); and the Meuse-Argonne offensive (Oct. 1 to 10, 1918, and Nov. 1 to 10, 1918). The operations which resulted in the capture of Blanc Mont and St. Etienne were construed to be included in the Meuse-Argonne offensive despite the fact that the operations were a part of the operations of the Fourth French Army, far to the west of the western limit of the American Meuse-Argonne sector and further that the work of the Second Division was continued by another American division. The operation which resulted in the capture of Hill 142, Bouresches, Bois de la Brigade de Marine, by the Marine brigade, assisted by Artillery, Engineers, etc., of the Second Division, and the capture of Vaux by the Third Brigade, Engineers and Artillery of the Second Division, were held to be local engagements rather than a major operation. The Second Division suffered about 9,000 casualties in the Chateau-Thierry sector.

In addition to the above major operations, Marine Corps personnel, other than that of the Fourth Brigade and Second Division, participated in the Champagne-Marne defensive, the Oise-Aisne offensive, and the Ypres-Lys offensive.

Chapter XVI.

WITH THE NAVY ON BOARD THE BATTLESHIPS AND CRUISERS.

WITH THE BRITISH GRAND FLEET.

Division 9 of the Atlantic Fleet, composed of the *New York* (flagship), *Wyoming*, *Florida*, and *Delaware*, was detailed for service with the British Grand Fleet, rendezvoused on November 24, 1917, in Lynnhaven Roads, Chesapeake Bay, and sailed for its destination the following day.

The division took the northern passage and was 13 days en route, 4 days of which were spent in holding its own against a 90-mile gale off the Newfoundland coast. On December 7, 1917, the American battleships anchored with the British Grand Fleet in Scapa Flow, Orkney Islands, after a rousing reception, and on December 26 were designated the Sixth Battle Squadron of the British Grand Fleet.

From the time of its arrival to November 29, 1918, this squadron, with the addition of the *Texas* in February, 1918, and the substitution of the *Arkansas* for the *Delaware* in July, 1918, operated with the British Grand Fleet, basing most of the time in Scapa Flow and the remainder of the time in the Firth of Forth (Rosyth). The squadron took its regular turn at convoy duty, patrol duty, target practice, and fleet exercises with all the other squadrons of the British Grand Fleet.

The squadron was at sea on an average of from 8 to 10 days each month and followed the procedure of the Grand Fleet in all respects, even going so far as to shift to the British method of signaling.

Maj. Nelson P. Vulte was division Marine officer of this division the entire time it was a unit of the British Grand Fleet.

With reference to the operations of this division the Secretary of the Navy in his annual report made the following statements:

> Assigned one of the two places of honor and importance in the battle line, this American division did its full share of the Grand Fleet's work, including patrol search for the enemy, protection of convoys, mining, and other forces, and, most important of all, in the repeated attempts to engage the German High Seas Fleet, for which the ships of the Grand Fleet were kept in the highest state of efficiency and readiness. Our battleships were attacked six times by submarines. On one occasion, off the Norwegian coast, four torpedoes were fired at the *Florida* and two at the *Delaware*, and at another time three were fired at the *New York*. Not one of our vessels was hit, and the only damage done was to the *New York*, which while leading the division into Pentland Firth, was rammed by a submerged submarine. Two blades of her propeller were broken off, but officers and crew were convinced that the blows from the propeller sank the U-boat.

SURRENDER OF THE GERMAN HIGH SEAS FLEET.

The American battleships occupied a prominent position in the north column of the Grand Fleet on the occasion of the surrender of the German High Seas Fleet, on November 21, 1918, off the mouth of the Firth of Forth and assisted in escorting it into that port where the German vessels were searched and later dispatched under guard to Scapa Flow, Orkney Islands, for internment.

The American vessels did not accompany the surrendered German war vessels to Scapa, but were detached from the British Grand Fleet on December 1, 1918, and sailed from Rosyth for Portland (Weymouth). The day after the surrender of the German Fleet the *Nevada*, which had been serving with Division 6 of the Atlantic Fleet in Bantry Bay, Ireland, joined Division 9, at Rosyth and proceeded with it to Portland.

AT CASTLETOWN BEREHAVEN, BANTRY BAY, IRELAND.

The Annual Report of the Secretary of the Navy makes the following remarks concerning Division 6:

> Division 6, composed of the *Utah* (flagship), *Nevada*, and *Oklahoma*, was based on Berehaven, Bantry Bay, Ireland, its principal duty being to protect our convoys from possible enemy raiders. This division made two trips into the Channel, escorting convoys when enemy submarines were reported in the vicinity.

Maj. Leon W. Hoyt was the division Marine officer of this division during its entire stay in European waters.

The *Nevada* joined the American battleships of Division 9 the day after the surrender of the German Fleet off Rosyth, near Edinburgh.

ESCORTING THE PRESIDENT INTO BREST.

Division 9 joined Division 6 at Portland Bill and both divisions left that port in time to assist the *Pennsylvania* in escorting President Wilson, on board the *George Washington*, into the harbor of Brest.

WELCOMED HOME BY NAVAL REVIEW.

On December 14, 1918, our battleships sailed from Brest for the United States, arrived off Ambrose Lightship the afternoon of Christmas Day, and the next morning steamed into New York Harbor where they were accorded a great demonstration. The naval review was followed by a land parade of all the returning officers, Bluejackets, and Marines.

THE ATLANTIC FLEET.

Col. (Brig. Gen.) John T. Myers was the fleet Marine officer of the Atlantic Fleet from before the outbreak of the war to August 23, 1918, being relieved on that date by Col. Frederic L. Bradman who continued on that duty until after the armistice.

Col. John F. McGill was force Marine officer of the Battleship Force and later force Marine officer of Battleship Force Two until August 14, 1918, when he was relieved by Maj. Harold C. Wirgman, who continued as such until the force was suspended in September, 1918.

Lieut. Col. James McE. Huey was force Marine officer of Battleship Force One from September 3, 1917 to December 29, 1917, when he was relieved by Maj. Edwin N. McClellan who continued as such until March 28, 1918, when the Marines were temporarily withdrawn from the force.

Maj. Richard H. Tebbs, jr., was force Marine officer of the Cruiser Force.

THE PACIFIC FLEET.

Col. Richard M. Cutts was fleet Marine officer of the Pacific Fleet from November 1, 1916 to October 14, 1918; and Lieut. Col. Charles B. Taylor from October 15, 1918 until after the armistice.

THE ASIATIC FLEET.

The Marines of the *Brooklyn*, flagship of the Asiatic Fleet, participated in the activities around Vladivostok, Siberia, in 1918.

In June, 1918, Vladivostok, and practically all of Siberia, was under the control of the Bolsheviki. The Bolsheviki, assisted by German and Austrian prisoners of war, were resisting the advance of the Czecho-Slovaks, who were trying to reach Vladivostok. In that city on June 29, 1918, there were approximately 12,000 well-organized Czecho-Slovaks, only about 2,500 of whom were armed or equipped. On the foregoing date the Czecho-Slovaks in the city took it over from the Bolsheviki after a three hour battle near its center, and on the afternoon of that day Rear Admiral Austin M. Knight, commander in chief of the Asiatic Fleet, ordered a detachment of American Marines ashore to guard the American consulate and to act as part of an Allied force composed of British, Japanese, Chinese, and Czecho-Slovaks, to patrol the city.

In July, 1918, Marines from the *Brooklyn* acted as guards over German and Austrian prisoners of war on Russian Island, about 5 miles from Vladivostok, while Marines from the same vessel constituted part of an Allied military force of American and British marines, Japanese and Chinese bluejackets, and Czecho-Slovak soldiers, which was organized to prevent a threatened strike and disorder among the workmen in the Russian navy yard at Vladivostok.

The *Albany* was at Vladivostok from April 2, 1919, until relieved by the *New Orleans* on July 25, 1919. Each of these ships, while they were anchored off Vladivostok, kept a small guard of Marines at the United States Naval radio station on Russian Island.

Col. Carl Gamborg-Andresen was fleet Marine officer of the Asiatic Fleet from August 25, 1915 to July 17, 1917; Col. Louis McC. Little from July 18, 1917 to April 25, 1918; and Col. Eli T. Fryer from that date until after the armistice.

Chapter XVII.

THE ACTIVITIES OF THE FIFTH BRIGADE.

THE FIFTH BRIGADE.

The units of the Fifth Brigade were never together as a brigade in France or at any time an element of a division, and for that reason its commanding general was assigned additional duty in the American Expeditionary Forces.

Brig. Gen. Eli K. Cole, the first commanding general of the Fifth Brigade, arrived in France on September 24, 1918, and proceeded to headquarters, Forty-first Division (First Depot Division), St. Aignan, France, reporting there October 1 to 4, 1918; on October 6, 1918, he joined the Second Division at Souain, France, as an observer and remained with the Second Division until October 26, 1918, when he left to report at St. Aignan. On October 28, 1918, Brig. Gen. Cole arrived at St. Aignan and assumed command of the Forty-first Division (First Depot Division). From December 27, 1918, to January 10, 1919, he commanded the First Replacement Depot. From January 12 to February 3, 1919, he commanded the American Embarkation Center at Le Mans, France. From February 23, 1919, to March 4, 1919, he commanded the Forwarding Camp at Le Mans, France. During the period March 5 to 21, 1919, Brig. Gen. Cole was occupied in inspecting the units of the Fifth Brigade. He left Tours, France, on March 31, arrived at Brest, France, same day, was detached from the American Expeditionary Forces on March 31, and sailed for the United States on the *Mauretania*, arriving at New York April 7, 1919.

Brig. Gen. Smedley D. Butler relieved Brig. Gen. Cole as commanding general of the brigade on April 9, 1919, and retained command until it was demobilized in August, 1919.

One of the most prominent and outstanding features of the American Expeditionary Forces was the administration of Pontanezen Camp at Brest, France, by Brig. Gen. Smedley D. Butler, United States Marine Corps, from October 6, 1918, to the latter part of July, 1919. The words of the citation conferring upon him the Army distinguished service medal describes in general terms the important work accomplished by Brig. Gen. Butler:

> Smedley D. Butler, brigadier general, United States Marine Corps. For exceptionally meritorious and distinguished services. He has commanded with ability and energy Pontanezen Camp at Brest during the time in which it has developed into the largest embarkation camp in the world. Confronted with problems of extraordinary magnitude in supervising the reception, entertainment, and departure of the large numbers of officers and soldiers passing through this camp, he has solved all with conspicuous success, performing services of the highest character for the American Expeditionary Forces.

Brig. Gen. Butler returned to the United States in command of the Fifth Brigade, on the *Siboney*, arriving at Hampton Roads, Va., on August 8, 1919.

Maj. William C. Wise was brigade adjutant from the date the brigade was organized to September 25, 1918; Maj. Charles D. Barrett relieved Maj. Wise and acted as adjutant until he went to the Fourth Brigade to relieve Lieut. Col. Ellis; Maj. Calvin B. Matthews was brigade adjutant from July 11, 1919, to the date the brigade was demobilized.

THE ELEVENTH REGIMENT.

Col. George Van Orden commanded the Eleventh Regiment during its entire existence.

The Eleventh Regiment was split up, its several units being spread all over France. Units of this regiment performed duty at various times at the following places: Brest, Tours, Montierchaume (Indre), Havre, Gievres (Loire-et-Cher), Marseilles, Toulon (B-du-Rhone), Miramas (B-du-Rhone), Issoudun (Indre), La Pallice, La Rochelle (Charante Inferieur), Mehun (Cher), St. Aignan-Noyers, Romorantin (Loire-et-Cher), Marans, Nevers, Aigrefeuille, Barmant, Somme, Chateauroux (Indre), Camp Covington (Camp Carret) near Marseilles, Paris (Headquarters Detachment, American Peace Commission).

The officers and men performed duties of various kinds, among such being: Post commanders, post and assistant post adjutants, personnel adjutants, regulating officers, assistant to the depot engineer, receiving officers, entertainment officers, assistant post chaplain, police officers, prison officers, camp guards, dock guards, commanding officers of troops, police sergeants, inspectors of the guard, district fire marshals, post welfare officers, district athletic officers, assistant provost marshals, fire patrol officers, fire marshals, transportation guard service, guard duty over prisoners, quartermaster property guard, interpreters, etc.

THE THIRTEENTH REGIMENT.

Col. Smedley D. Butler commanded the Thirteenth Regiment from the date of its organization until November 19, 1918, on which date Lieut. Col. Douglas C. McDougal assumed command and remained its commanding officer until it was demobilized.

Like the Eleventh Regiment, the units of this regiment performed duty in the various posts in the Services of Supply, among such places being Brest, Bordeaux, St. Nazaire, La Rochelle, La Pallice, Rochefort, Montoir, Bassens (Gironde), Sursol (Gironde), Casino-de-Lilas (Bordeaux), La Teste (Gironde), Beau Desert (Gironde), Nantes, St. Sulpice (Gironde), Savenay, St. Loubes (Gironde), Lormont, Carbon Blanc, Grange Neuve, Genicart, Croix d'Hins, La Baule, Isle of Ste. Anne (Nantes), Pen Houet, Usine Brulee.

The officers and men performed duties of various kinds, among such being provost guard, hospital center guard, camp guard, railroad transportation officers, commanding dock guard, dock guard, unloading ships, erecting tents at Pontanezen Barracks, military police, warehouse guards, convoying of railroad trains, special guards for shipments of commissary supplies, assistants to camp commander at Pontanezen Camp, prison guards, assisting thousands of convalescent and sick soldiers who disembarked from the *Leviathan* to get to Camp Pontanezen, inspector general's department, base section No. 1, stockade guard, traffic police, motor transportation convoy guard, dock guard secret service, segregation camp, and railway patrol.

THE FIFTH BRIGADE MACHINE-GUN BATTALION.

Maj. Ernest A. Perkins commanded the Fifth Brigade Machine Gun Battalion from the date of its organization until November 4, 1918; Capt. Franklin A. Hart from that date until November 12, 1918; and from November 12, 1918, to date of demobilization Maj. Allen H. Turnage was the commanding officer.

This battalion performed duty at Camp Pontanezen during its entire stay in France.

Chapter XVIII.

CASUALTIES.

During the period of the World War the Marine Corps personnel suffered casualties in actual battle in France with the American Expeditionary Forces (Second Division and Aviation); in Aviation while operating as part of the naval service in France; and in the West Indies in operations against the bandits of Santo Domingo.

MARINE CORPS CASUALTIES.

Marine Corps deaths in the American Expeditionary Forces, as obtained from Marine Corps records on January 14, 1920, are divided as follows:

Character.	Officers.	Enlisted men.	Total.
Killed in action	45	1,420	1,465
Died of wounds received in action	30	961	991
Died of accident	3	24	27
Died of disease	14	255	269
Other causes	1	11	12
Total	93	2,671	2,764

The following is a summary of the casualties sustained by the Fourth Brigade of Marines from March 15 to November 11, 1918, as published in General Orders, No. 66, Second Division, American Expeditionary Forces, dated July 2, 1919:

Fourth Brigade casualties.

	Killed.		Died of wounds.		Missing.		Wounded severely.		Wounded slightly.		Wounded, degree undetermined.		Gassed.		Total.	
	Officers.	Men.	Officers.	Men.	Officers.	Men.	Officers.	Men.	Officers.	Men.	Officers.	Men.	Officers.	Men.	Officers.	Men.
Toulon sector (Verdun), Mar. 15–May 13	....	12	2	44	....	....	....	16	6	46	....	89	4	287	12	494
Aisne defensive and Chateau-Thierry sector, May 31–July 9, 1918	25	724	9	304	....	33	3	322	20	656	46	2,123	9	436	112	4,598
Aisne-Marne offensive, July 18–25, 1918	10	156	5	104	....	66	9	250	20	489	30	911	2	39	76	2,015
Marbache sector, Aug. 9–22	....	1	....	1	....	....	....	2	....	8	....	3	....	....	....	15
St. Mihiel offensive, Sept. 12–Sept. 16	3	57	....	86	....	11	3	124	6	334	4	287	....	4	16	903
Meuse-Argonne (Champagne), Oct. 1–Oct. 10	9	320	8	155	....	31	23	380	43	862	6	480	6	141	95	2,369
Meuse-Argonne offensive, Nov. 1–11	8	189	1	59	....	20	7	222	20	467	6	206	3	55	45	1,218
Total	55	1,459	25	753	....	161	45	1,316	115	2,862	92	4,099	24	962	356	11 612

AVIATION CASUALTIES.

The following table shows the casualties sustained by the Marine Aviation forces between April 6, 1917, and November 11, 1918:

Character.	Officers.	Enlisted men.	Total.
Killed in action	2		2
Died of wounds received in action	1		1
Died of accident	6	6	12
Died of disease	1	25	26
Died of other causes	1		1
Wounded in action	2	3	5
Total	13	34	47

TOTAL MARINE CORPS DEATHS.

From April 6, 1917, to September 10, 1919, 131 officers and 3,489 enlisted men died, a total of 3,620 Marine Corps deaths from all causes.

CASUALTIES IN THE DOMINICAN REPUBLIC.

During the period between April 6, 1917, and November 11, 1918, one officer was wounded in action, four enlisted men were killed in action, and thirteen wounded in action, in the Dominican Republic in operations against bandits.

LOST ON THE CYCLOPS.

Two Marines died when the *Cyclops* was lost at sea.

CASUALTIES OF NAVAL PERSONNEL.

Of the 60 naval medical officers, 12 naval dental officers, and 500 enlisted men of the Medical Corps of the Navy serving with the Marines in the American Expeditionary Force, 1 commissioned officer, and 12 enlisted men were killed; 8 commissioned officers and 101 enlisted men were wounded or gassed.

Chapter XIX.

CITATIONS OF MARINE ORGANIZATIONS—DAYS IN FRANCE—ARTILLERY CAPTURED—PRISONERS CAPTURED—KILOMETERS ADVANCED—DECORATIONS AWARDED.

CITATIONS IN FRENCH ARMY ORDERS.

The French Army recognized the splendid work of the Fifth and Sixth Regiments of Marines by citing them three times in Army orders for achievements in the Chateau-Thierry sector, the Aisne-Marne, and the Meuse-Argonne (Champagne). The Sixth Machine Gun Battalion was similarly cited for its work in the Chateau-Thierry sector and the Aisne-Marne, and the Fourth Brigade for its work in the Chateau-Thierry sector.

Information was received in January, 1920, that the War Department had accepted the award of the French fourragère in the colors of the ribbon of the Croix de Guerre for several Army organizations and the three units of the Fourth Brigade.

Prior to this date the only American organizations which had received permission to accept or wear the French fourragère were three sections of the ambulance service and one aero squadron, all of which were temporary organizations and have now been demobilized.

DAYS IN FRANCE.

A Marine Corps unit arrived in France with the first expedition of of American troops. From June 26, 1917, to November 11, 1918, Marines were in Europe with the American Expeditionary Forces a total of 504 days, of which 66 days were in active sectors and 71 in quiet sectors.

ARTILLERY CAPTURED BY SECOND DIVISION.

The commanding general, Second Division, under date of December 30 1918, reported to General Headquarters, American Expeditionary Forces, the following data with reference to artillery and machine guns captured:

Sector.	Heavy artillery.	Light artillery.	Trench mortars.	Machine guns.	Antitank guns.
Verdun	....	....	....	....	....
Chateau-Thierry	....	(1)	12	119	....
Soissons	9	66	2	200	....
Marbache sector	....	....	....	....	....
St. Mihiel	60	61	....	122	....
Blanc Mont	5	37	27	409	8
Meuse-Argonne	....	[2] 105	17	500	....
Total	74	269	58	1,350	8

[1] A small number of light artillery was captured, but no count made.

[2] On account of tho rapid advance, for a total of about 29 kilometers, during which time these guns, in position and along the roads, were overrun and left behind, it was impossible to make an accurate check of of them ,and therefore the figures report guns both heavy and light. They were taken from reports of subordinate commanders made at the time. Rifles were not counted.

PRISONERS CAPTURED.

The Second Division captured 12,026 prisoners, which is 19.07 per cent of the total prisoners captured by the entire American Expeditionary Forces.

KILOMETERS ADVANCED.

The Second Division advanced 60 kilometers against the enemy.

DECORATIONS AWARDED MARINES.

The following number of decorations were awarded Marines during the war:

Medals of honor (Army)	5
Distinguished-service medals (Army)	8
Distinguished-service crosses (Army)	363
Distinguished-service order (British)	1
Croix de guerre (French)	1,237
Legion of honor (French)	19
Medaille militaire (French)	10
Belgian decorations	10
Chinese decorations	1
Italian decorations	9
Montenegrin decorations	4
Portuguese decorations	1
Total	1,668

The above number of Distinguished-service crosses (Army) awarded includes 42 awarded to Navy Medical Corps personnel, 2 to Y. M. C. A. personnel, and 2 to French officers serving with Marines.

The above number of Croix de guerre (French) awarded includes 82 awarded to Navy Medical Corps and Navy Dental Corps personnel, and 3 to Navy Chaplains serving with Marines.

One Navy Chaplain was awarded a Legion of Honor (French), but this is not included in the above.

Chapter XX.

RIFLE PRACTICE—RIFLE AND PISTOL COMPETITIONS.

RIFLE PRACTICE.

In recent years the Marine Corps has devoted a great deal of time and energy to rifle practice, believing that one of the first requirements of a soldier is to know how to shoot. During the period of the war target practice was given special attention, and in 1918 it was announced that no enlisted men would be sent overseas who had not qualified as marksman or better. This announcement created even greater interest than before in target practice among the enlisted personnel and gratifying results were obtained on all rifle ranges. The Marines that arrived in France were educated riflemen, but despite that fact rifle ranges of some character were established and used in every spot of France and Germany where the Marines remained long enough to make it feasible and practicable to do so.

The percentage of marksmanship qualifications of the enlisted personnel of the Marine Corps on various dates, in the American Expeditionary Forces and in the United States was as follows:

Date and place.	Percentage.
Entire Marine Corps, Apr. 6, 1917	48.0
Marines of American Expeditionary Forces, Nov. 11, 1918	68.1
Entire Marine Corps, Nov. 30, 1918	67.1
Entire Marine Corps, Mar. 1, 1919	66.0
Marines of American Expeditionary Forces, July 1, 1919	81.6

The number of marksmanship qualifications of the last six years in the Marine Corps was as follows:

Classification.	1914	1915	1916	1917	1918	1919
Expert rifleman	596	883	1,287	1,709	6,019	7,851
Sharpshooter	2,749	2,536	1,984	2,373	8,933	10,642
Marksmen	757	1,471	2,594	6,011	14,826	21,918
Total qualified	4,102	4,890	5,865	10,093	29,778	40,411
Percentage	0.415	0.493	0.591	0.379	0.670	0.828

RIFLE AND PISTOL COMPETITIONS.

During the period of the war the Marine Corps rifle teams, teams representing Marine Corps units, and teams partly composed of Marines, engaged in seven important competitions.

(*a*) The National Rifle Association and the national matches held at Camp Perry, Ohio, in 1918. Marines won the following National

Rifle Association matches: Members' match, 300-yard rapid-fire match, Wimbleton cup match, Marine Corps match, President's match, and the grand aggregate; and took second place in the Leech cup match and the 200-yard rapid-fire match. Of the national matches, Marines won the national team match and United States Service match; took sixth, ninth, eleventh, twelfth, twenty-sixth, thirtieth, sixty-seventh, sixty-eighth, sixty-ninth, and seventy-first places in the national individual match; took second place in the national individual pistol match; and three Marines were among the first hundred of the individual pistol match in which there were 942 shooters entered.

(*b*) The National Rifle Association and the national matches held at Caldwell, N. J., in 1919. The members of the 1919 Marine Corps rifle team squad made a splendid showing in the rifle matches, outclassing their military and civilian competitors in almost every match held. The Marines won 13 matches out of the 16 in which they were entered; civilian riflemen took 2 events; and the Cavalry 1. Marines won the following matches: Company team, enlisted men's team, member's, Marine Corps cup, rapid fire, regimental team, veteran team, two-man team, President's, grand aggregate, national individual, United Service, and the national team. Four Marines were on the American Expeditionary Forces team, which took second place.

(*c*) The twenty-sixth annual Sea Girt interstate tournament held at Sea Girt, N. J., in 1919. The Marines won 14 of the 18 matches in which they were entered, winning the following matches: Hayes, Meany, Spencer, two-man team (New Jersey), Wingate, Libbey, all-comers expert, Cruikshank trophy, Rogers trophy, Sadler trophy, Dryden trophy, McAlpin trophy, Roe all-comers long range, and Sea Girt championship.

(*d*) The American Expeditionary Forces rifle, pistol, and musketry competition, held on the d'Avours range at Le Mans, France, in May, 1919. The first three places in the individual rifle competition were won by Marines; a Marine won the individual pistol match; the Fifth Regiment of Marines stood first in the regimental standing, followed by the Thirteenth, Sixth, and Eleventh Regiments in seventh, eighth, and eleventh places in the order mentioned; a Marine won first place in the individual automatic rifle competition. The Second Division led all other divisions.

(*e*) The Inter-Allied championships held on the d'Avours range at Le Mans, France, in July, 1919. The American Expeditionary Forces team, on which were four Marines, defeated all nations. A Marine took second place in the individual rifle match.

(*f*) Third Army championship (Amaroc shoot) held on the rifle range at Wehr, Germany, under the auspices of the Third Division in June, 1919. The Marines and the Second Division won most of the honors in this competition.

(*g*) A special Inter-Allied rifle competition for five-men teams on a 300-meter range near Paris, France. France won and America was second. Two Marines were on the American team.

Chapter XXI.

AVIATION.

STRENGTH AND DISTRIBUTION.

On April 6, 1917, the Marine section of naval aviation consisting of five officers and 30 enlisted men, was stationed at the naval air station, Pensacola, Fla., as part of the complement of that station.

During April, May, and June, 1917, the Marine aviation section was transferred to a combination land and water station for Marine fliers at the navy yard, Philadelphia, Pa., and the training of personnel for land flying began. The official designation of this organization was the Marine Aeronautic Company. Training in observation balloons was done in addition to the heavier-than-air work.

On October 12, 1917, this Marine Aeronautic Company, then consisting of 34 officers and 330 enlisted men, was divided into the First Aviation Squadron, consisting of 24 officers and 237 enlisted men, and the First Marine Aeronautic Company, consisting of 10 officers and 93 enlisted men.

On October 14, 1917, the First Marine Aeronautic Company was transferred to Cape May, N. J., and took over the naval air station at that place.

On December 7, 1917, the First Marine Aeronautic Company, then consisting of 12 officers and 133 enlisted men, was ordered to Naval Base 13, Ponta Delgada, Azores, arriving there on January 21, 1918. This company was the first completely equipped American aviation unit to leave the United States for service in the war. This organization operated an antisubmarine patrol station of 10 *R-6* seaplanes, 2 *N-9* seaplanes, and later 6 *HS-2-L* flying boats until the station was ordered abandoned on January 24, 1919, when it was ordered to return to the United States, arriving at the Marine flying field, Miami, Fla., March 15, 1919. Maj. Francis T. Evans was in command from January 9 to July 18, 1918, and Maj. David L. S. Brewster from July 19, 1918, to January 20, 1919.

On October 17, 1917, the First Aviation Squadron was transferred from the Marine flying field, navy yard, Philadelphia, Pa., to the Army training field at Mineola, Long Island, where instruction and training were carried on in land flying. On December 31, 1917, this organization was transferred to Gerstner Field, Lake Charles, La., for advanced training.

In March, 1918, the Marine flying field, Miami, Fla., was established and on March 31, 1918, the First Aviation Squadron was transferred to that field from Lake Charles, La.

Four Marine squadrons of land-fighting planes and a headquarters company were organized to operate under the Navy as the Day Wing of the Northern Bombing Group, in northern France, which operated in the Dunkirk area against German submarines and their bases at Ostend, Zeebrugge, and Bruges.

On July 13, 1918, the first Marine aviation force, consisting of Squadrons A, B, C, and Headquarters Company, left Miami, Fla., and embarked on board the *De Kalb* at New York City for France, July 18, 1918. This organization consisted of 107 officers and 654 enlisted men, and when Squadron D joined in October, 1918, it consisted of 149 officers and 842 enlisted men. On July 30, 1918, the Day Wing disembarked at Brest, France, and proceeded to its aerodromes between Calais and Dunkirk, where they established camp and prepared the aerodromes for use. The personnel of the Day Wing was completely organized and ready for service two weeks after their arrival in France. Part of the planes and equipment of this organization arrived at Pauillac, France, before the organization reached France on July 30, 1918. On September 28, 1918, one plane was delivered to the Marine Day Wing. On October 5, 1918, Squadron D of the Day Wing, consisting of 42 officers and 188 enlisted men, arrived at Le Franc aerodrome, completing the four squadrons of the Day Wing. During the month of October additional planes were delivered to the Day Wing.

In order to prevent the personnel, who were completely trained and ready for action when they reached the front, August 2, 1918, from getting badly out of practice, the commanding officer, Day Wing, requested permission from the British aviation forces in the vicinity to be allowed to assign certain Marine pilots to operate with their squadrons until the Marine planes were delivered. As many Marine pilots as could be accommodated were operating with British squadrons until the end of the war, and were highly complimented by the British officers. The Day Wing, carried out 14 independent raids far behind the enemy lines, did considerable damage, and brought back valuable information. The organization participated actively and creditably in both offensives on the Flanders front. It was learned after the armistice that one raid resulted in the death of 60 enemy officers and 300 enlisted men. A feat worthy of mention was performed by Marine Corps pilots. A French regiment was cut off by the enemy near Stadenburg. It was decided to attempt to feed them by aeroplane. Marine Corps pilots loaded up with food and flew low over this isolated regiment and successfully dropped 2,600 pounds of food to them in the face of heavy fire from artillery, machine guns, and rifles. This process was continued for two days until the regiment was extricated. The number of enemy planes brought down by Marine pilots, bombs dropped, food dropped, and other facts of a statistical nature are given elsewhere. Three pilots were killed or died of wounds received in action, two of them being shot down over the enemy's lines.

Maj. Alfred A. Cunningham commanded the Day Wing from the date of its organization to December 7, 1918, except the period August 1 to 7, 1918, during which time Maj. Roy D. Geiger was in command.

While in Europe the Marine fliers served with Squadrons 213 (pursuit squadron), 217, and 218 (bombing squadrons), Royal Flying Corps of England; and with pursuit, observation, and bombing squadrons of the French Flying Corps.

In February, 1918, the Marine aviation section of 8 officers and 40 enlisted men was organized and stationed at the naval air station, Miami, Fla. The personnel of this section was later increased and served at that station throughout the war, taking over the deep-sea scouting of that station. Capt. Thomas R. Shearer was in command during the entire time.

STRENGTH AT BEGINNING AND END OF WAR.

The strength of Marine aviation on April 6, 1917, and on November 11, 1918, was as follows:

APRIL 6, 1917.

Commissioned officers	4
Warrant officer	1
Enlisted men	30
Total	35

NOVEMBER 11, 1918.

Commissioned officers	250
Warrant officers	32
Enlisted men	2,180
Total	2,462

STATIONS.

The following are the stations at which Marine aviators operated, showing whether they operated independently, with the Navy, or with the Army:

Independently.	Navy.	Army.
Marine flying field, Miami, Fla. Balloon Company, Marine Barracks, Quantico, Va. Naval Base No. 13, Azores. Marine flying field, Philadelphia, Pa. Naval air station, Cape May, N. J.	Marine section naval air station, Miami, Fla. Day Wing, Northern Bombing Group, France. Naval air station, Pensacola, Fla.	Roosevelt Field, Mineola, Long Island. Gerstner Field, Lake Charles, La. Army balloon schools at St. Louis, Mo., and Omaha, Nebr.

PLANES OPERATED BY MARINES.

The number of planes operated by Marine aviators at Pensacola, Fla., on April 6, 1917, was four, and the type, AH Curtiss.

On November 11, 1918, the following planes were operated by the personnel of Marine Aviation:

Marine flying field, Miami, Fla.:	
De Haviland 4's, Curtiss JN's, Thomas-Morse scouts, and M-1 defense planes	118
Naval air station, Miami, Fla.:	
HS-1-L, and HS-2-L flying boats, and R-6 Curtiss seaplanes	24
Balloon Company, Quantico, Va.:	
N-9 and R-6 seaplanes	3
Caquot and kite balloons	4
Naval base No. 13, Azores:	
R-6 and N-9 seaplanes and HS-2-L, and HS-1-L flying boats	18
Northern Bombing Group, France:	
De Haviland 4's and 9's	177

At other times the following planes were operated by Marine pilots:

At Philadelphia, Pa.:	
Curtiss JN's	6
Curtiss R-6's	2
Curtiss N-9's	2
At Roosevelt field, Mineola, Long Island:	
Army land planes	12
At Gerstner field, Lake Charles, La.:	
Army land planes	12
At Cape May naval air station:	
R-6 and N-9 seaplanes	8
Total planes operated by Marine pilots	386

TRAINING OF OFFICERS AND ENLISTED MEN.

During the war the Marine Corps selected and trained its own flyers and mechanics, and had its own aviation field and equipment. At the Massachusetts Institute of Technology, Boston, Mass., enlisted Marines selected as promising flying material and given the rank of gunnery sergeant, took a 10 weeks' course in groundwork, and about 80 men a month were graduated. After ground graduation they did their actual flying at the Marine flying field, Miami, Fla. This course embraced preliminary, aerobatic, and formation flying, bombing, gunnery, and reconnaissance work, including photographing. Upon qualifying they were commissioned as second lieutenants in the Marine Corp Reserve Flying Corps. Marine flying candidates were all enlisted Marines, of superior physique, weighing from 135 to 165 pounds, and with at least two years' college or university study to their credit. The age limits were 19 to 39 years. Marine Corps mechanics, riggers, and armorers were trained at the Marine Corps section of the naval school for mechanics, Great Lakes Training Station, Chicago, Ill., the course covering eight weeks, and at a similar school in aviation mechanics at San Diego, Calif.

In December, 1917, 2 Marine officers and 10 enlisted men were sent to the Army balloon school at St. Louis, Mo., and later to Omaha, Nebr., for training.

AVIATION STATISTICS.

Marine squadrons overseas in France	5
Total officers in France	165
Total enlisted men in France	895
Marine officers serving with Army Air Service, American Expeditionary Forces	6
Overseas, outside of France: 1 squadron, 12 officers, 135 enlisted men, Naval Base No. 13, Ponta Delgada, Azores.	
Officers completely trained ready for overseas aviation duty in the United States on Nov. 11, 1918	100
Enlisted men completely trained ready for overseas aviation duty in the United States on Nov. 11, 1918	1,150
Number of squadrons and companies in United States (Nov. 11, 1918)	3
Total officers in United States on Nov. 11, 1918	100
Total enlisted men in United States Nov. 11, 1918	1,150
Total number of cadets under training (at all times)	225
Cadets completely trained (in all three branches) as bombers, chasse pilots, and seaplane fliers, total commissioned from Oct. 1, 1918, to date	175
Raids participated in by Marine fliers serving with the British and French	43
Total number of bombing raids completely Marine	14
Total pounds of bombs dropped	52,000

Number of food raids	5
Pounds of food dropped	2,600
Number of enemy aircraft accounted for officially	12

RETURN OF MARINE AVIATORS FROM EUROPE.

Early in December, 1918, the Day Wing received orders to return to the United States, and embarked on December 6, 1918, on board the *Mercury* at St. Nazaire, France, arriving at Newport News, Va., December 21, 1918.

The First Marine Aeronautic Company returned from the Azores in March, 1919, arriving at the Marine flying field, Miami, Fla., on March 15, 1919.

Chapter XXII.

MARINE CORPS RESERVE.

On April 6, 1917, the strength of the Marine Corps Reserve, all classes, was 36; the enlisted strength of the Naval Militia, Marine Corps Branch, of the various States, was 928.

On April 1, 1917, the Naval Militia, Marine Corps Branch, attained its highest strength, 1,046. There was no recruiting for the Naval Militia, Marine Corps Branch, after April 1, 1917, and in consequence, the enlisted personnel of that branch showed a steady decrease subsequent to that date, owing to discharges and rejections.

On July 1, 1918, the Naval Militia, Marine Corps Branch, then the National Naval Volunteers, Marine Corps Branch, consolidated with the Marine Corps Reserve in pursuance with the provisions of an act of Congress, approved July 1, 1918, and in consequence thereof became members of class 2, Marine Corps Reserve.

On November 16, 1918, the Marine Corps Reserve attained its highest strength, 6,773.

STRENGTH ON NOVEMBER 11, 1918.

The following table shows strength of Marine Corps Reserve on active duty on November 11, 1918:

Majors	7
Captains	33
First lieutenants	63
Second lieutenants	360
Total commissioned officers	463
Marine gunners	27
Quartermaster clerks	2
Pay clerks	4
Total warrant officers	33
Enlisted men	6,483
Female reservists	277
Total enlisted personnel	6,760

Strength of the Marine Corps Reserve.

	Total men, all active.	Total men, inactive.	Women, active.	Women, inactive.	Total.
1917.					
Apr. 1	35				35
May 1	424				424
June 1	885				885
July 1	1,096				1,096
Aug. 1	1,167				1,167
Sept. 1	1,186				1,186
Oct. 1	1,210				1,210
Nov. 1	1,202				1,202
Dec. 1	1,341				1,341
1918.					
Jan. 1	1,531				1,531
Feb. 1	2,514				2,514
Mar. 1	4,106				4,106
Apr. 1	4,745				4,745
May 1	4,780				4,780
June 1	4,950				4,950
July 1	5,211				5,211
Aug. 1	6,378				6,378
Sept. 1	6,453		31		6,484
Oct. 1	6,402		145		6,547
Nov. 1	6,467		240		6,707
Dec. 1	6,440	42	269		6,751
1919.					
Jan. 1	5,820	588	260		6,668
Feb. 1	5,022	1,301	255		6,578
Mar. 1	4,392	1,910	246		6,548
Apr. 1	3,555	2,684	234		6,473
May 1	3,064	3,041	229		6,334
June 1	2,871	3,179	238		6,288
July 1	2,410	3,502	226		6,138
Aug. 1	2,186	3,694		201	6,081

Chapter XXIII.

RETURN OF MARINES FROM EUROPE—PARADES IN THE UNITED STATES.

RETURN OF SECOND DIVISION AND FOURTH BRIGADE.

A great many Marines were returned from Europe gradually and in small detachments from the date the armistice became operative.

The commanding general of the Second Division and his staff, headquarters of the Fourth Brigade, the Fifth Regiment, and the Second Battalion of the Sixth Regiment arrived in the United States on board the *George Washington* on August 3, 1919; the remainder of the Sixth Regiment arrived in the United States early in August, 1919, on board the *Rinjdam* and the *Wilhelmina;* the Sixth Machine Gun Battalion arrived in the United States on board the *Santa Paula* on August 5, 1919.

RETURN OF THE FIFTH BRIGADE.

The Fifth Brigade Headquarters, the Thirteenth Regiment (less Company B), and the Fifth Brigade Machine Gun Battalion arrived in the United States on board the *Siboney* on August 8, 1919. Company B of the Thirteenth Regiment arrived on the *Mercury* on August 12, 1919. The Eleventh Regiment arrived in the United States on board the *Orizaba* on August 9, 1919.

All the above Marine organizations and individuals were returned to the naval service soon after arrival in the United States.

COMPOSITE REGIMENT, THIRD ARMY.

The company of Marines and battalion commander (major) and staff, forming a part of the Composite Regiment, Third Army, returned to the United States on board the *Leviathan* on September 8, 1919, and were returned to the naval service in September, 1919.

The colonel commanding the Composite Regiment in a letter dated September 21, 1919, commended the battalion commander and staff, the commanding officer of the company, and " the lieutenants for their loyalty and attention to details, and noncommissioned officers and men for their soldierly appearance, high standard of morale, and discipline," concluding with these words:

> The Composite Regiment paraded as escort to the general of the Army, in London, Paris, New York, and Washington, D. C. The regiment has been favorably commended. This is entirely due to the loyalty, energy, and attention to duty of the officers and individual soldiers in the regiment; and in this the Marine Corps representatives deserve a large share.

SCHLESWIG-HOLSTEIN BATTALION.

With the return of the above Marine organizations, all Marines of the American Expeditionary Forces were out of Europe with the exception of a few individuals, and the Fifteenth Separate Battalion, consisting of 26 officers and about 700 enlisted men, which was retained for duty in France for possible operations in connection with the Schleswig-Holstein plebiscite.

This battalion, under command of Maj. Charles F. B. Price, was organized at Pontanezen Camp, Brest, in July, 1919, from personnel of the Fourth and Fifth Brigades and the Twelfth Separate Battalion. The battalion designation was changed on August 15 from "Provisional Battalion, U. S. Marines," to the "Fifteenth Separate Battalion."

The Battalion rendered honors to Gen. Pershing on September 1, 1919, upon his departure from France at Brest and on the same day was inspected by Marshal Foch, who commended the Battalion on its splendid appearance.

Leaving Brest on the *Mercury*, September 3, 1919, the Battalion six days later arrived at Bordeaux to take part in the ceremony of laying a foundation for a monument commemorating the entrance of the United States into the World War, at Pointe de Grave, near Bordeaux, France. It then returned to Brest and in December went on board the *Henderson*, then at that port. The *Henderson*, with the Battalion on board, sailed from Brest, arriving at Philadelphia on December 23, 1919, after a 16-day trip. On December 30, 1919, the battalion arrived at Quantico, Va.

RETURN OF AVIATION UNITS.

Information regarding the return of Marine Corps aviation units from Europe will be found in chapter XXI.

PARADES IN THE UNITED STATES.

On August 8, 1919, the Fourth Brigade, as a part of the Second Division, paraded in New York City. Maj. Gen. John A. Lejeune, with many Marine officers on his staff, was in command.

On August 8, 1919, the Fourth Brigade of Marines was transferred to the naval service upon its arrival at Quantico, Va.

On August 12, 1919, the Fourth Brigade of Marines, then a part of the naval service, was reviewed by the President of the United States in a parade at Washington, D. C. Brig. Gen. Wendell C. Neville was in command.

A company of Marines and a battalion commander, as a part of the Third Army Composite Regiment, paraded in New York and in Washington, D. C., as escort to Gen. Pershing. The First Division also formed a part of these parades.

Chapter XXIV.

DEMOBILIZATION.

Immediately upon the armistice becoming operative on November 11, 1918, the question of demobilization became one of paramount importance. It was necessary that plans be at once put into effect providing for the release from service at the earliest possible date of duration-of-war men and reservists. While authority existed to hold such men for several months after the ratification of the peace treaty, the popular demand for the return and discharge of all who could be spared was only natural. Parents, relatives, and friends could see no necessity for the keeping of their loved ones in the service after actual fighting ceased.

The problem before the Marine Corps was serious, as a wholesale reduction at that time would have seriously crippled its efficiency. Therefore, on November 20, 1918, in Marine Corps Orders No. 56, orders were issued to the service stating that it was the desire of headquarters to release those members of the Marine Corps Reserve and those men of the Regular Service, who enlisted for the duration of the war, who wished to complete their education, or who had urgent family and business interests demanding immediate and personal attention. Thus demobilization to a limited extent was begun nine days after the signing of the armistice.

On May 1, 1919, it became necessary, owing to the demands of the service and the reduction of the enlisted personnel, temporarily to limit the privilege of discharge to men whose release was necessary for urgent financial dependency reasons.

Following the approval of the act of July 11, 1919, which act provided sufficient funds to sustain the corps only at an average enlisted strength of 27,400 men, with corresponding officers, Marine Corps Orders No. 42, July 12, 1919, were published, establishing demobilization centers and promulgating detailed instructions for the complete demobilization. Under this plan duration-of-war men were discharged as rapidly as the exigencies of the service permitted. Those men who were serving in the Tropics who were eligible for discharge and desired their release were returned to the United States as rapidly as practicable, and orders were issued for their discharge and awaited them at the time of joining a Marine barracks, thus minimizing delay in allowing them to go home. By the latter part of December, 1919, practically all of the duration-of-war personnel had been discharged.

In August, 1919, the Fourth and Fifth Brigades of Marines, which had been serving with the Army in France, were returned to the Marine barracks, Quantico, Va., and the naval operating base, Hampton Roads, Va., respectively, at which places the demobilization of these two organizations was effected. This undertaking was by far the largest and most important of its kind that had ever confronted the Marine Corps, but due to the coordination of the various departments interested, and the far-reaching and clearly defined instructions issued in advance the demobilization of these units was effected in a remarkably short time, being completed on August 13, 1919, and in a manner bringing satisfaction to the men discharged, and reflecting to the credit of the corps. The success of these efforts is evidenced by the following statement of demobilization:

Discharged or transferred to inactive status:
Fourth Brigade, 6,677 enlisted men.
Fifth Brigade, 6,671 enlisted men.

In addition to the enlisted men released there were also about two hundred officers of the two brigades who were either discharged or transferred to an inactive status. Due to the diligent cooperation of those concerned the demobilization was carried out with a degree of success far beyond expectations.

MARINES ARE WELCOMED HOME BY THE NAVY.

Acting Secretary Roosevelt August 11, 1919, sent to all ships and stations of the United States Navy the following special order:

The Fourth Brigade of the Marine Corps, consisting of the Fifth Regiment, Sixth Regiment, and the Sixth Machine Gun Battalion, and the Fifth Brigade of the Marine Corps, consisting of the Eleventh Regiment, Thirteenth Regiment, and Fifth Brigade Machine Gun Battalion, have returned from service in Europe and have reverted from the Army to their status in the United States Navy. At this time I wish, on behalf of the naval service, to welcome them back and express to the officers and men of these organizations the very deep appreciation of the Navy for their splendid services while with the Army during the war.

Beginning with the first expeditionary forces which left the United States in June, 1917, over 30,000 officers and men of the Marine Corps have been sent to France. The Fourth Brigade, as a part of the immortal Second Division, has been engaged in all of the principal operations of the war. Their record speaks for itself. The Fifth Brigade, going to France later, furnished many splendidly trained replacements for the Fourth Brigade and performed arduous tasks according to tradition.

The entire Navy welcomes them home.

FRANKLIN D. ROOSEVELT,
Acting Secretary of the Navy.

THE SECRETARY OF WAR PRAISES THE FOURTH BRIGADE.

WAR DEPARTMENT, *August 12, 1919.*

Hon. JOSEPHUS DANIELS,
Secretary of the Navy.

MY DEAR MR. SECRETARY: In the process of demobilization, the Marine Brigade, which by the President's order became a part of the American Expeditionary Forces and was thus a part of the forces under the control of the War Department and under the command of Gen. Pershing, has now been returned to this country, detached from the Army, and restored to the control of the Navy Department.

I can not permit this heroic force to terminate its association with the Army without expressing to you, and through you to the officers and men of the Marine Corps, the deep sentiment of the War Department and of the Army toward it. The whole history of the Brigade in France is one of conspicuous service; when it was finally incorporated into the Second Division of the American Army it had early an opportunity to give a heroic demonstration of the unconquerable tenacity and dauntless courage of American soldiers. From then on in successive, almost continuous, battles the Marine Brigade and the division of which it was a part fought sternly and successfully until victory was obtained for the Allied Armies. Throughout this long contest the Marines, both by their valor and their tragic losses, heroically sustained, added an imperishable chapter to the history of America's participation in the World War.

On behalf of the Army I congratulate the Navy Department, the Major General commanding the Marines, those who have been instrumental in the formation and training of this splendid organization, and the officers and men of the organization itself.

Cordially, yours,

NEWTON D. BAKER.

In reply Acting Secretary Roosevelt said:

NAVY DEPARTMENT, *August 13, 1919.*

Hon. NEWTON D. BAKER,
Secretary of War, Washington, D. C.

DEAR MR. SECRETARY: Your very cordial letter and the tribute it bore to the Fourth Brigade of Marines was received with pleasure and deepest appreciation. The heroism of the Marines and the Regulars in the famous Second Division, and their sacrifices, have endeared them to all Americans, and it is with very pardonable pride that we welcome them back to the Navy.

The spirit of cordial cooperation between the Army and the Navy was never better manifested than in the participation of these Marines in the great battles in France under the command of Gen. Pershing as a part of the United States Army, and shoulder to shoulder with units of the Regular Army. It is with extreme gratification that we can look back upon this unbroken cooperation between our two departments that started at the time the first Navy ship carried troops to France and continued uninterruptedly through to the end.

On behalf of Secretary Daniels, the Commandant of the Marine Corps, the officers and men of that organization, I wish to thank you for the sentiments expressed in your letter and convey to you our appreciation of the heroism of the officers and men of the Army who with the Marines made the Second Division one of the greatest fighting organizations the world has ever known.

It is very gratifying in our pride over the achievements of the Marines, to know that that pride is shared by the War Department and your warm approbation of their conduct as a part of the Army will be treasured by the Corps as well as by the individuals.

Sincerely, yours,

FRANKLIN D. ROOSEVELT,
Acting Secretary of the Navy.

Chapter XXV.

OFFICE OF THE MAJOR GENERAL COMMANDANT—ADJUTANT AND INSPECTOR'S DEPARTMENT.

Maj. Gen. Commandant George Barnett was the Major General Commandant of the United States Marine Corps during the entire period of the World War. Originally appointed on February 25, 1914, he was reappointed on February 25, 1918, for a second term of four years.

On September 29, 1918, Maj. Gen. Commandant George Barnett, accompanied by Brig. Gen. Charles L. McCawley, sailed from New York on board the *Leviathan*, arriving at Brest, France, October 7, 1918. The object of the visit of the Major General Commandant to France was an inspection of all the Marines serving with the American Expeditionary Forces, but he fell a victim to the influenza epidemic which prevented him from carrying out his plans. He departed from Paris, December 7, 1918, sailed from Brest, December 9, 1918, on board the *Leviathan*, and arrived in the United States, December 16, 1918.

Brig. Gen. John A. Lejeune was the Assistant to the Major General Commandant from December 14, 1914 to September 26, 1917, when he was transferred to Quantico, Va., to command the Marine barracks. Brig. Gen. Charles G. Long relieved Brig. Gen. Lejeune and has acted as Assistant to the Major General Commandant from that date to the present.

Activities directly under the office of the Major General Commandant such as personnel, target practice, and aviation, were carried on efficiently during the war. The Planning Section was established on December 24, 1918.

Brig. Gen. Charles H. Lauchheimer was the Adjutant and Inspector of the United States Marine Corps, with station at Headquarters, during the World War. He became seriously ill, was admitted to the hospital on July 10, 1919, where he died on January 14, 1920. Col. Henry C. Haines assumed the duties of Acting Adjutant and Inspector on August 3, 1919, and upon the death of Brigadier-General Lauchheimer was appointed the Adjutant and Inspector.

A great amount of additional work was caused by the large increases and by the war, but the personnel of the Adjutant and Inspector's Department performed their duties with efficient satisfaction.

Chapter XXVI.

PAYMASTER'S DEPARTMENT.

Brig. Gen. George Richards was the Paymaster of the United States Marine Corps, stationed at Headquarters, during the entire period of the war.

STRENGTH AND DISTRIBUTION.

The commissioned, warranted, appointed, and enlisted personnel of the paymaster's department, at the beginning of the war, consisted of:

Permanent commissioned paymasters	6
Officers of the grade of captain detailed for four years	3
Officers appointed as special disbursing agents under Revised Statutes 3614	4
Permanent pay clerks	9
Civil force	3
Enlisted men	51
Total force	76

The above force was distributed as shown in the following table:

Pay stations.	Personnel attached.						
	Number and rank commissioned officers.	Pay clerks.	Quartermaster sergeants.	Sergeants.	Corporals.	Privates.	Total.
Established offices in United States.							
Headquarters, Washington, D. C.	1 brigadier general; 1 lieutenant colonel 1 major, 1 first lieutenant.	4 [1] 3	4	5	2	6	28
San Francisco, Calif	1 major	1	1	2	1	1	7
New York, N. Y	1 captain	1	3	3	2	4	14
Outside of United States.							
Peking, China	1 captain [2]		1				2
Virgin Islands	do.[2]		1				2
Port au Prince, Haiti	1 first lieutenant		1	1	1	1	5
Cape Haitien Haiti	1 captain [2]		1		1		3
Santo Domingo, Dominican Republic.	1 captain	1	1		1		4
Santiago, Dominican Republic.	do	1	2	1		3	8
Managua, Nicaragua	1 second lieutenant[2]		1				2
Unassigned		1					1
Total	13	12	16	12	8	15	76

[1] Civil. [2] Special disbursing agent.

The commissioned force of the paymaster's department reached a maximum strength of 47 officers, including special disbursing agents, when the latest detail to the department from the line was made, and as then constituted, consisted of:

Permanently commissioned paymasters	5
Officers detailed from the prewar line	9
Temporary officers (eight former pay clerks and four temporary line officers)	12
Officers of the reserve force	16
Officers appointed as special disbursing agents	5
Permanent pay clerk	1
Temporary pay clerks	58
Pay clerks of the reserve force	6
Enlisted men	501
Total authorized force	613

The above force which was the maximum authorized complement of the paymaster's department, was distributed as follows:

Pay stations.	Number and rank commissioned officers.	Pay clerks.	Quartermaster-sergeants.	Sergeants.	Corporals.	Privates, first class.	Privates.	Total.
Established offices in United States.								
Headquarters, Washington, D. C.[1]	1 brigadier general, 4 majors, 5 captains.	22	27	37	55	22	67	240
New York, N. Y. (established May, 1908).	2 majors, 1 captain	2	5	5	5	2	8	30
Philadelphia, Pa. (established, May, 1917).	1 colonel, 1 captain	2	2	2	2	1	3	14
Quantico, Va. (established June, 1917).	1 lieutenant colonel, 1 major, 2 captains.	4	5	8	8	3	13	45
Atlanta, Ga. (established June, 1917).	1 captain	3	3	3	4	2	6	22
San Francisco, Calif. (established March, 1901).	do	2	2	3	3		3	14
Outside United States.								
Expeditionary Forces in France.	4 majors, 12 captains	20	34	34	31	9	39	183
Seventh Regiment, Cuba	1 captain	1	1	1	2		1	7
First Regiment, Cuba	do	1	1	1	2		1	7
San Domingo, Dominican Republic.	do	1	1	2	2		1	8
Santiago, Dominican Republic.	do	1	1	1	2		1	7
Port au Prince, Haiti	1 special disbursing agent.	1	1	1	1			5
Cape Haitien, Haiti	do		1		1		1	4
St. Thomas, Virgin Islands	do	1	1					3
Managua, Nicaragua	do		1					2
Peking, China	do		1					2
Attached to posts in United States and elsewhere.	1 captain	4	15					20
Total	47	65	102	98	118	39	144	613

[1] Authorized complement but never filled. Maximum number employed in paymaster's office at any time subsequent to Apr. 6, 1917, was approximately 208 of all grades and ranks.

HOW WAR INCREASES WERE MET.

The enlargement of the paymaster's department to meet war conditions at the beginning of the war, and until the enlisted strength was raised to 75,500, was effected in the following manner:

1. By the temporary appointment and advancement to the grade of captain, pursuant to act of May 22, 1919, of eight of the permanent pay clerks of the department.

2. By the temporary advancement of enlisted men of experience and long service under this department to the grade of pay clerk.

3. The new enlisted clerical personnel was partly obtained from enlisted men who had previously been employed as pay roll clerks at shore stations and aboard ships of the Navy, and from men enlisted and enrolled from civil life with clerical experience outside. The men obtained were detailed in the regularly established offices, and there formed into classes for instruction in their duties, the commissioned officers and senior clerks being used as instructors for this purpose. In addition to the above, and to the end of creating a proper spirit and morale, and bringing about a better understanding throughout the department of its aims and purposes, a series of lectures by the Paymaster and subordinate officers was delivered at headquarters, and afterwards published and distributed to the entire personnel of the department. Later on, in order to meet the further increased demand for clerical assistance, a school for the instruction of men in paymaster's department work was established at the Marine barracks, Parris Island, S. C. This was, however, in addition to the system of instruction previously instituted in the permanently established offices. The school was of considerable value in that it aided in the selection and assignment of men (recruits) with previous clerical experience to duty in the paymaster's department. The demand for clerks for both home and overseas service, however, was so great for some time before the close of the war, that it was not possible at any time to keep the men under instruction in the school or in the offices for sufficient length of time to complete the prescribed course that had been laid out for them. As a consequence, many men had to be sent out with but a meager idea of the duties they were to perform. The clerical forces of the permanent offices, therefore, finally became so drained of experienced clerks and stenographers taken away to supply the demand for expeditionary and overseas forces that it became necessary to enlist or enroll women to perform these duties.

MONEY EXPENDED.

The amount of money expended for pay and allowances for each month from April, 1917, to December, 1918, follows:

Months.	Officers.	Enlisted men.	Total.
1917.			
April	$143,698.37	357,398.60	501,094.97
May	176,742.19	413,019.82	589,762.01
June	201,977.79	565,677.22	767,655.01
July	220,884.23	989,495.53	1,210,379.76
August	280,038.19	1,196,082.09	1,476,120.28
September	269,160.66	1,137,790.16	1,406,950.82
October	332,859.49	1,244,965.55	1,577,825.04
November	378,458.48	1,266,138.57	1,644,597.05
December	350,098.46	1,353,510.60	1,703,609.06
1918.			
January	328,517.81	1,322,724.98	1,651,242.79
February	337,407.35	1,441,099.32	1,778,506.67
March	369,298.24	1,321,933.64	1,691,231.88
April	373,395.58	1,313,210.75	1,686,606.33
May	378,551,77	1,815,309.63	2,193,861.40
June	385,042.99	1,856,344.35	2,241,387.34
July	441,838.15	2,124,134.76	2,565,972.91
August	515,273.26	2,436,318.86	2,951,592.12
September	572,721.39	2,322,089.79	2,894,811.18
October	522,995.22	2,503,312.62	3,026,307.84
November	648,615.69	2,597,542.71	3,246,158.40
December	597,142.61	2,545,932.41	3,143,075.02
Totals	7,821,715.92	32,124,031.96	39,948,747.88

ADDITIONAL DUTIES DURING WAR.

The duties of the paymaster's department during the war were greatly enlarged and made more complicated and difficult: (1) By reason of the enactment of the war risk insurance act of October 6, 1917. The work connected with family allotments and war risk insurance created by this act was of such magnitude as to require the establishment of a separate administrative section under a commissioned officer to handle the voluminous correspondence, keep the records, and make proper audit of these items in the accounts involved. It was also found necessary in order to facilitate the work of this section, that a liaison group of clerks be kept in the War Risk Bureau. (2) By reason of the taking over of the payment of all Marine Corps allotments, as the deputy of the Navy allotment officer. (3) By reason of the necessity of having to pay many men on affidavits without proper records; service record books and other papers pertaining to their accounts having been lost or destroyed by operation of war or other accidental circumstances. (4) By reason of the large number of wounded men, some of whom were scattered in various hospitals throughout France, and others of whom were returned to the United States without due notice to military authorities, and sent to both naval, military, and civil hospitals at widely scattered points throughout the States. In but a few of these cases were there any records on which full and accurate payments could be made. Hence a system of emergency, or casual payments, as they were called, was established both in France and in the United States. The absence of records in these cases was not the worst feature however, but the absence of any information whatever, as to the whereabouts of the men made it at first impossible to locate some of them and effect regular payments. However, after the first few months' experience with the handling of payments to the wounded, a system was devised by which most of those returning to the States were immediately reported and prompt payments were thereafter made. A similar system of emergency or casual payments to wounded men was adopted by the department in France, but wounded men in France were evacuated so frequently from one hospital to another, that no system of reporting was practicable. Each hospital there had to be visited in person by a paymaster at least once a month and such wounded Marines as were found, had to be paid on their own representations a sum sufficient to meet their immediate needs. Under such a system some necessarily went without pay for some time, while others more fortunate in meeting a paymaster at frequent intervals, received at times more money than was properly due them. (5) By reason of the enactment of February 24, 1919, providing a gratuity of $60 to all persons in the military and naval forces of the United States, who were discharged under honorable conditions at any time subsequent to April 6, 1917. This law necessitated the establishment at headquarters of a claims section, whose sole duty was to settle the twenty or thirty thousand supplementary claims created by this act and the act of February 28, 1919, increasing the amount of travel allowance to 5 cents per mile to all enlisted men discharged subsequent to November 11, 1918.

NEW PAY ROLL.

In addition to the above, it became necessary to adopt a new pay roll suitable for preparation on the typewriter and so arranged as to make it adaptable for use as a combination pay and muster roll, should this be deemed necessary. This roll was prescribed and put into use in the midst of the war without much confusion, and it is understood resulted in a saving of much clerical labor to the organization commanders.

OFFICE OF THE CHIEF PAYMASTER, U. S. MARINES, FRANCE.

In obedience to orders dated October 2, 1917, and in compliance with provisions of G. H. Q. General Orders No. 38(2), September 17, 1917, the "Office of the Chief Paymaster, U. S. Marines, France," was established in Paris, France, on October 5, 1917. Maj. Davis B. Wills was Chief Paymaster, U. S. Marines, France, from that date until the office was abolished in August, 1919.

Chapter XXVII.

QUARTERMASTER'S DEPARTMENT.

Brig. Gen. Charles L. McCawley, was the Quartermaster of the United States Marine Corps, with station at Headquarters, during the entire war. Brig. Gen. McCawley, in company with the Major General Commandant, sailed from the United States on board the *Leviathan* September 29, 1918, arriving at Brest, France, October 7, 1918. After an extended visit to the Marines as an observer Brig. Gen. McCawley sailed from Brest on board the *De Kalb* December 9, 1918, arriving in the United States, December 16, 1918.

Upon the increase of the Marine Corps from 17,400 to 30,000 and later to 75,500, it became necessary to increase the commissioned, warrant, and enlisted personnel of the quartermaster's department, in order that it might successfully meet the heavy demands made upon it by the war and by the large increase in strength.

The increase in the commissioned personnel was made largely by the promotion of experienced quartermaster clerks and quartermaster sergeants to commissioned rank, and the vacancies in the grade of quartermaster sergeant were filled, as far as possible, from selected enlisted men.

Later a school for the instruction of quartermaster sergeants was organized at Marine barracks, Norfolk, and three classes were graduated therefrom.

It was necessary to make large increases in the personnel at headquarters to handle the question of supply, transportation, construction, and finance. The table below shows the strength in the office of the quartermaster on January 1, 1917, and June 30, 1918:

Grade.	Jan. 1, 1917.	June 30, 1918.
Commissioned officers	4	9
Warrant officers		12
Special assistant		1
Technical engineer		1
Clerical force:		
Civilians	10	8
Enlisted (regular)	43	53
Enrolled (reservists)		111
Total	57	195

ADDITIONAL STOREHOUSES AND COOPERATION WITH GOVERNMENT AGENCIES.

Due to changes in the method of purchasing rations, caused by existing conditions, it became necessary to establish commissary storehouses at San Francisco, Charleston, and Baltimore, at which to maintain reserve supply stores. The Baltimore storehouse was later moved to Philadelphia. At the beginning of the war the Marine Corps had in its depots at Philadelphia, Pa., and San Francisco, Calif., a small surplus stock, which had been accumulated from the regular appropriations, sufficient to outfit 8,500 men; consequently when the war was declared against Germany and the corps was increased, first to 1,323 officers and 30,000 men (act May 22, 1917), and secondly to 3,341 officers and 75,500 men (act July 1, 1918), it was necessary to provide simultaneously clothing, equipage, food, and shelter for these men. The question of shelter is discussed in the following pages. In connection with the purchase of these supplies it must be remembered that the Army, Navy, and Allies were in the market for similar articles, and in order to determine supply and allocate demands the President first appointed a Council of National Defense, which was later superseded by the War Industries Board. The quartermaster's department had representatives in daily attendance at the various meetings of the committees of the Council of National Defense, and later the War Industries Board. All of the above-mentioned supplies, as far as practicable, were purchased in the usual manner, by the bid and tender plan, only those articles on which no bids were received or those controlled by the War Industries Board being allocated.

There was installed in the office of the quartermaster a "follow-up" system, where record was made of all orders, contracts, purchase orders, and requests for transfer of supplies from other departments. The function of this section was to see that the supplies were delivered in accordance with contract obligations and trace delinquent deliveries. The section has proven its value, and satisfactory deliveries have been obtained, with few exceptions. From these records the quartermaster has available at all times the status of all outstanding orders as well as a concise record of completed contracts.

CANTONMENTS.

To furnish accommodations for the increased personnel, cantonments on a large scale were built at Quantico, Va., and Parris Island, S. C., and on a smaller scale at Mare Island, Calif. This work was expeditiously handled and afforded suitable temporary accommodations during the war.

DEPOT OF SUPPLIES, PHILADELPHIA, PA.

During the period of the war the depot outfitted and equipped 36 expeditionary units for service in France and the West Indies, and over 31,000,000 pounds of various kinds of supplies were shipped on Government bills of lading. The depot departments were so organized that it was only necessary to expand each division of the office forces and increase the number of employees and machines in the manufacturing departments in order to meet the increased demands during the war. The personnel of the depot on June 30, 1919, was as follows: Thirteen commissioned officers, 7 warrant officers, 2 civilians, 102 enlisted men of the regular service, 21 reservists, and 1,095 other employees of all classes, making a total personnel of 1,240.

DEPOT OF SUPPLIES, SAN FRANCISCO, CALIF.

The activities of this depot were increased during the war by the greater number of recruits to be outfitted on the west coast, and by the establishment of the subsistence branch of the depot at San Francisco. This depot has supplied all posts on the west coast, and furnished the supplies for the troops in the Orient.

DEPOT OF SUPPLIES, CHARLESTON, S. C.

This depot was established soon after the declaration of war for the purpose of supplying all posts south of Norfolk, including the West Indies. The storage facilities consist of 7 warehouses and a total floor space of 124,778 square feet. A total of about 14,287 tons of stores were shipped from this depot during the fiscal year 1919, these stores being valued at approximately $12,000,000; during the same period approximately 18,000 tons of stores were received, at an estimated value of $15,000,000. Practically all shipments of supplies of every description for troops in the West Indies are made from this depot.

EXPENDITURES.

The expenditures of the quartermaster's department for the fiscal years ending June 30ths, 1917, 1918, and 1919, were as shown below, exclusive of appropriations for public works, under the Navy Department, from which figures the enormous increase in the activities and responsibilities of this department, caused by the war and by the increase in strength, is evident:

Subhead.	1917	1918	1919
Provisions	$1,612,908 30	$6,725,893.05	$10,287,965.53
Clothing	2,173,501.59	11,123,760 36	20,275,456.01
Fuel	248,606.82	590,120.91	989,573.08
Military stores	1,520,289.39	6,371,978.10	13,952,476.49
Camps of instruction	31,871.01	30,015 83	....
Transportation and recruiting	620,667.75	1,514,657.77	3,064,099.21
Repairs of barracks	216,715.56	3,754,241.58	5,883,065.69
Forage	75,018.94	161,614 81	163,132.90
Commutation of quarters	164,497.24	402,402.51	363,484.53
Contingent	983,984.91	4,864,825.25	8,674,269.61
Expenditures under appropriation "Reserve supplies, U. S. M. C."	....	....	2,510,527.44
Purchases under second deficiency act from United States Army	....	....	772,540.00
Total maintenance quartermasters' department U. S. M. C.	7,648,061.54	35,540,440.17	66,936,590.49

INDEX.

A.

B.

C.

H.

I.

J.

K.

L

O.

T.

U.

V.

ADDITIONAL COPIES
OF THIS PUBLICATION MAY BE PROCURED FROM
THE SUPERINTENDENT OF DOCUMENTS
GOVERNMENT PRINTING OFFICE
WASHINGTON, D. C.
AT
10 CENTS PER COPY

Made in United States
North Haven, CT
10 November 2026

22084524R00065